Media Affirmation of the Warning Call Sounded by *Cell Phones: Invisible Hazards in the Wireless Age*

"Carlo and Schram have done a commendable job of pulling together information that will help consumers make informed decisions. And their final message is lucid: 'The big picture is becoming disturbingly clear: There is a definite risk that the radiation plume that emanates from a cell phone antenna can cause cancer and other health problems.' Carlo and Schram have raised serious questions. It's now up to science and industry to provide honest answers."

—*San Francisco Chronicle*

"Carlo did research into the safety of wireless devices and in his book traces the cell phone industry's eagerness to sell its product and how government agencies have failed to act to protect us or even tell us about evidence of health risks."

—*Denver Post*

"Cell-phone companies contend the fears are unfounded but, savvy marketers that they are, most are quietly introducing more efficient—and therefore lower-radiation—phones. . . .

The cell-phone industry's best known naysayer is George Carlo, a scientist who once headed a controversial five-year, $25 million industry-sponsored study of possible radiation hazards. 'I just don't want people to put these phones to the sides of their heads,' says Carlo, who this month published a scathing book about his findings, called *Cell Phones: Invisible Hazards in the Wireless Age*. Carlo maintains that his data show plenty of cause for concern; he uses a hands-free headset that keeps his frequently busy mobile phone away from his brain."

—*Time*

"Carlo said he felt the industry, specifically the trade group Cellular Telecommunications Industry Association was 'spinning the science in a way we felt was dangerous to consumers' and the government was not doing all it should, either.

That's why he decided to write the book with Washington-based journalist Martin Schram that was recently published by Carroll & Graf Publishers.

'The industry made a commitment to research and when the findings were not what they wanted, they decided to go in another direction,' he said.

He claimed there is a 'powerful media manipulation campaign that is confusing consumers. There is no question but there is a health risk.'"

—*Buffalo News*

Dr. George Carlo and Martin Schram

Cell Phones

Invisible Hazards in the Wireless Age

An Insider's Alarming Discoveries about Cancer and Genetic Damage

With a New Preface by the Authors

Carroll & Graf Publishers, Inc.
New York

CELL PHONES: INVISIBLE HAZARDS IN THE WIRELESS AGE

Carroll & Graf Publishers
An Imprint of Avalon Publishing Group Inc.
245 West 17th Street, 11th Floor
New York, NY 10011

First Carroll & Graf cloth edition 2001
First Carroll & Graf trade paperback edition 2002

Library of Congress Cataloging-in-Publication Data is available.

ISBN: 0-7867-0960-X

Printed in the United States of America
Distributed by Publishers Group West

9 8 7 6 5 4 3

Table of Contents

PART TWO

PART THREE

PART FOUR

DEDICATION

"To my family and friends who encouraged me to do the right thing, and then supported me when I did. And to my grandson Ethan, in the hope that this work will make his world a safer place."

—G.C.

"For Steven and Christopher, and all others in the first generation of the wireless age."

—M.S.

An Update For the Paperback Edition of *Cell Phones: Invisible Hazards in the Wireless Age*

It has been almost one year since the publication of our book which comprises Cell Phones: Invisible Hazards in the Wireless Age, an urgent wake-up call for responsible action by government and industry. Since then, there has indeed been much action—but unfortunately, more action than progress.

The U.S. government provided an official model of how it is possible to produce action without progress. In response to a request from members of Congress, auditors for the U.S. Congress's General Accounting Office looked at the unusual arrangement by which the U.S. Food and Drug Administration allows the wireless communications industry to finance and control research into whether the industry's devices pose any threat to public health. But the congressional auditors' report steered a cautious middle ground—stopping short of recommending federal funding for much-needed independent research.

Indeed, the congressional auditors' report also failed to fulfill the specific requirement of Sen. Joseph Lieberman (D-Conn.), who, in requesting the study along with Rep. Edward Markey (D-Mass.), had asked the auditors "to clarify what [health] risks may or may not exist" for cell phone users. The auditors presented no clarifying conclusions about health risks. They focused only minimally upon laboratory experiments that were detailed in this book (and which were explained in detail to them by Dr. George Carlo, whom they interviewed in the course of their study). That was especially unfortunate because those lab experiments found that radiation from cell phones caused genetic changes in human blood cells and animal tissue—changes that cancer experts have repeatedly called a diagnostic marker of "high risk" for developing tumors. In August, Senator Lieberman asked the congressional auditors to continue monitoring the FDA's partnership with the industry it is regulating.

The result of this action without progress is that U.S. policy in this public health issue remains in marked contrast to the policies in Britain and France, where governments fund and run cell phone safety research. Indeed, British government officials issued a recommendation that all cell phones should carry warning labels alerting buyers that children shouldn't use the phones because their skulls are more readily penetrated by cell phone radiation.

While the government's executive and legislative branches have been slow to act, the judicial system has been focusing on mobile phones and public health. Class-action lawsuits and personal injury lawsuits have been filed in a number of states. Some lawsuits seek compensation from cell phone manufacturers and providers for health problems allegedly suffered by frequent cell phone users. And importantly, some of the suits also seek funds to provide consumers with headsets that will enable people to keep the antenna safely away from their heads while using cell phones.

While publication of our book sparked an initial flurry of national and international media attention on the potential health risks posed by cell phones, scientists and epidemiologists throughout the world continued their research and published further results. Some, but not all, of those studies seemed to confirm the findings and concerns raised in this book. In Germany, epidemiologists published findings showing a tripling of the risk of melanoma of the eye associated with cell phone usage. In Sweden, scientists found a doubling of brain cancer risk in cell phone users who had used their phones for ten years.

However, there were also three statistical studies, widely reported in some major U.S. news media, that unfortunately conveyed false assurances to cell phone users—due to the fact that the studies contained fundamental sampling flaws. Those shortcomings were candidly noted in the fine print of the most prestigious of those studies, done under the auspices of the National Cancer Institute and published in the New England Journal of Medicine. Yet most in the media overlooked those shortcomings, as they prominently reported as reassuring news the studies—two in the United States, one in Denmark—that showed no link between cell phone usage and cancer. One shortcoming in the studies was that they dealt with patients who had used cell phones for only short periods of time (less than three years, in the two U.S. studies) and thus did not address what might happen to those who use cell phones frequently, over a decade or more. As the National Cancer Institute researchers' article responsibly noted: "The most important limitation of our study is its limited precision for assessing the risks after a potential induction period of more than several years or among people with very high levels of daily or cumulative use." The second, crucial shortcoming of the three studies was that the vast majority of tumors in the patients studied were located in interior regions of the skull that couldn't have been reached by cell phone radiation. That is because cell phone radiation that emanates from the antenna penetrates only two inches inside adult skulls; the researchers needed to study the types of tumors that appear near the surface, along the sides of the heads of people who use cell phones. Again, it was the National Cancer Institute authors who candidly noted this limitation, conceding that "a much larger sample would be required" to properly detect risks from cell phone radiation.

In other words: The statistical studies only proved that tumors couldn't be reached by cell phone radiation weren't caused by cell phone usage. That's hardly reassuring. Yet news reports that overlooked those caveats unfortunately may have lulled cell phone users into thinking they did not have to take the most basic, commonsense precaution called for in this book: Just as it is impossible for consumers to buy a new car without getting a seat belt, it should be impossible to buy a cell phone without getting a headset, or ear-piece, to keep the antenna harmlessly away from the head.

And that brings us to the one reassuring development that did occur this year. It came not from government regulation or congressional mandate, but from a voluntary decision by one major company in the wireless telecommunications industry that has been explored in our book and has been cited in an increasing number of lawsuits. In September, AT&T announced that it will give free headsets to all its customers.

Will people who begin using cell phones as children or teens—and use them frequently in their 20s and 30s—be at high risk for developing brain cancer in their 40s and 50s? So far no one in the industry, science, or government can answer that question. And that is why the warning call we sound in our book remains an urgent call for action by us all.

Dr. George Carlo and Martin Schram
October 2001
Washington, D.C.

Introduction

It was the Orwellian year of 1984. Some Americans were wondering if Walter Mondale would really be able to capture the presidency by defeating the incumbent, Ronald Reagan. Some were wondering if the Chicago Cubs would really be able to capture the National League championship by defeating the San Diego Padres. But almost nobody was wondering about the potential cancer-causing risk of what was surely the year's most technologically, socially, and culturally transforming event—the first mass marketing of an instrument called a cellular telephone. It was a wireless marvel by which people could reach into their pockets, then reach out and touch someone—anywhere in the world.

At first, it seemed like just another toy for people with money; it was rather expensive to buy and use each month. There had been no premarket testing of this radiation-emitting device that people would use while pressing it against their heads; after all, it was viewed as more of a luxury and curiosity than a necessity. But after a slow start, just-plain people everywhere began to feel this was one instrument they just had to have. The telephone manufacturers quickly discovered that clever and costly advertising campaigns would produce wondrous payoffs—for the manufacturers, and the carriers that provided the telephone service.

By 1993 there were 15 million cellular telephones in use in the United States alone. Still the U.S. government had taken no action and apparently given little thought to whether or not this radiation-emitting instrument that people were pressing to their skulls would cause any adverse health effects.

In the past few years, the technological and cultural revolution that is the Wireless Age has virtually exploded around the world. By the year 2000 there were more than 103 million mobile phones in the United States—and more than 500 million mobile phones in use worldwide. This instrument that was once satirical fodder for comedians poking fun at the boring rich had become a neo-necessity and plaything of the masses. People who may have barely been able to pay their rent and grocery bills were walking down the streets, sitting on buses, and even in restaurants with their mobile phones pressed against their heads, chattering away even while they appeared to be by themselves.

Meanwhile, the wireless revolution continues growing exponentially. Visit any public building, college classroom, courthouse, or commuter train, and look around: You'll see people using not just wireless phones but also wireless laptop computers and miniature palm tops. What you won't see are the microwaves that are crisscrossing a confined space where a number of people who are not even using these instruments are bombarded by these waves. What is happening with these waves is not unlike the scene from the movie *The Saint* in which infrared burglar detector beams crisscross a room. Only the waves from wireless instruments cannot be seen. And whether or not they will cause any damage, near-term or long-term, is unknown. Society was not really prepared for the wireless revolution—nor was politics, or science. There was no policy in place anywhere in the world to safeguard the public as these instruments of unknown potential risk were suddenly available to all.

• • •

It was 1993 when the cellular telephone industry was rocked by the first wave of damaging news. CNN's *Larry King Live* devoted an entire show to a lawsuit that was filed on behalf of a cancer victim against the cell phone industry. To ease public concerns and governmental pressures, the industry turned to Dr. George Carlo, a public-health scientist, to head a $25 million research and surveillance program that, in the words of the industry, would assure the public that cell phones were safe. In the early years, when Carlo initially found no cancer or other health problems, his relations with the industry were basically tranquil. But later, he implemented a series of studies using newly designed exposure systems; they produced findings that raised red flags of warning that cell phone radiation could indeed lead to the development of brain tumors, other cancers, or other adverse health effects. The industry reacted by treating him as an enemy—to be ostracized in public and discredited in private.

Yet as the calendar approached 2001, the U.S. government still had taken no action to advise the public of the scientific evidence that had generated these red-flag warnings of cancer and other adverse health effects.

One good way to grasp all that has happened—and mainly, all that hasn't—is to think about the cellular telephone from a totally different perspective: If a mobile phone (which facilitates oral communication) were something that was taken orally, like a pill or capsule, the government would have required that it be premarket tested to assure its safety before it ever reached the hands of consumers. And if the testing had turned up evidence such as the findings produced by a number of studies, including those in Carlo's program, cell phones probably would have been withheld from sale to the public until modifications were made—such as mandatory use of a headset, at a minimum—to assure that use of the product was safe. At the very least, there would have been a regulation requiring that warning labels be placed on the telephones. (It is worth noting, for instance, that studies by the Swedish scientist Dr. Kjell Hansson Mild found that cell phone users experienced headaches and dizziness. Those symptoms appear frequently on warning labels of many pharmaceutical products, at the insistence of the U.S. Food and Drug Administration [FDA].)

But because cell phones are held alongside the brain rather than swallowed like a capsule, they were never premarket tested. And there will be no post-market recall. But meanwhile, the government has been slow to act and the industry is not about to voluntarily make major design changes or recommend the use of headsets—because that could be considered an industry admission that there is some sort of problem. And that could put the industry at risk in future class-action lawsuits.

• • •

This book tells the story of the controversy over the cellular telephone—and what it means for every man, woman, and child—by recounting the scientific investigations that were led by one man. Carlo began his quest in 1993 as the industry's designated insider on the case. But as his team began pursuing an independent path, the industry began treating him as an outsider within.

This book follows two paths simultaneously: First, it follows the science of cell phone radiation. Second, it follows the politics involved, which have been every bit as complex—for this was an effort to determine whether products of a burgeoning industry, suddenly the fastest-growing industry in the world, can cause cancer and other health problems to the people it serves.

Because of the book's unusually layered storyline, following both the science and the politics of the cell phone controversy, it is written not in one voice but in two. To provide a straightforward narrative, the report is presented in the third person; to provide much-needed perspective and context in the presentation of the events, the account is based both on Carlo's notes, files, and recollections, and also upon additional reporting by co-author Martin Schram. In the end, Carlo was responsible for the recollections of events that transpired and for all analysis of the science. There are a number of occasions where additional reporting and detached insight have made it clear that mistakes were made by many of the players in the probe of cell phones and public health. And those errors—by industry executives, government officials, prominent scientists, and to be sure, by Carlo himself—are painstakingly reported in this book with no punches pulled.

There are also many occasions where Carlo speaks directly to the reader in the first person; these are presented in *italic* type. In these instances, Carlo is recalling an occurrence of significance and/or is providing a clear and simple explanation of a complex scientific finding. On other occasions, Carlo offers first- person accounts that are not really about scientific findings, but just provide one man's tale of what it was like to be caught up in the vortex of these powerful scientific and political storms. For readers unfamiliar with some of the scientific matters involved, there is a glossary at the back of the book.

Two organizations that play a pivotal role in the story this book tells decided that their officials would not be interviewed for it. Cellular Telecommunications Industry Association President, Thomas E. Wheeler declined through a spokesperson to be interviewed. Three officials of the U.S. government's Food and Drug Administration (FDA) declined through a spokesperson to be interviewed unless they were provided in advance with the questions that would be asked of them. Martin Schram told the FDA spokesperson that is something he had never been required to do in more than three decades of reporting in Washington, D.C. during which he has interviewed presidents, cabinet members (including those who have overseen the FDA), their subordinates, and members of Congress. He explained to the FDA that this would be journalistically improper for him to do.

While this chronicle of the cell phone health controversy is told from Carlo's perspective, it most surely is not about one person. It is about 103 million people who use cell phones in the United States—and 500 million who use them around the world. They are the ones who deserve to know what the problem really is, and what solutions are available to help them communicate safely in the wireless age.

PART ONE

CHAPTER ONE

WHEN SCIENCE MEETS POLITICS

THE STREETS AND sidewalks of Long Beach, California, were bake-oven hot on this mid-June day in 1999, but that was nothing compared to the heat that was being generated inside the air-conditioned comfort of the Hyatt Regency Hotel. There, in a conference room, more than 100 scientists and dozens of trade-press journalists from five continents were attending a "State of the Science" colloquium, convened to discuss the public-health impact of cell phones. The audience was listening with more than just scientific interest. For at the podium was the organizer of this colloquium, Dr. George Carlo, a public-health scientist with graying hair and a grayer beard, who many in the hall used to refer to as the cell phone industry's "hired hand" (but always behind his back, of course!). Now Carlo was sounding the one warning the industry executives who had funded his research least wanted to hear.

"It is very clear to me that everyone is doing their job—and the consumers are not being protected," Carlo declared ruefully. He outlined new evidence indicating that cell phones may indeed cause cancer and other health damage to consumers. Months earlier, he had given the findings to the cellular telecommunications industry's top Washington lobbyist, Thomas E. Wheeler—the man who hired him six years earlier to run the industry's science research program. But the industry had kept Carlo's troubling new findings carefully under wraps. Now, Carlo stunned most of the audience by calling upon the industry's top officials to tell the public everything they knew about the health risks posed by mobile phones—and to develop an entirely new standard for the amount of radio wave emissions that can safely emanate from these instruments which people everywhere hold against their heads. The old standard, he said, was based upon old data, old science, and old theories that were now invalid—perhaps dangerously so. His words rang alarm bells throughout the industry.

Tom Wheeler had not bothered to fly out from Washington, D.C., just to hear this too-public warning issued by the scientist he'd personally brought into his cell phone inner circle in 1993. Back then, the industry's strategy seemed to be a masterstroke of science-veneered damage control; it would take a scientist to keep the scientific community in check, keep the government regulators at bay, and keep the cell phone consumers blithely buying—by assuring one and all that cell phones are safe.

At first, back in the early and mid-1990s, Carlo, a medical professor of epidemiology and a cautious, deliberate researcher, had been quite comfortable issuing public assurances that scientific research had found no health risks in radiation from cellular telephones. But then, as he followed the science, he came up with new research evidence that cast serious doubt on those early studies. Early in 1999 the new evidence raised cause for serious concern about health risks and made additional research an urgent imperative. Carlo had gone to the offices of the president of the Cellular Telecommunications Industry Association (CTIA) and told Wheeler about the existence of credible new evidence of health risks. When he added he could no longer say publicly that the research showed no health risks from cell phone radiation, the industry lobbyist moved quickly to distance himself from the scientist who was the bearer of bad news. And he did it in a mover-and-shaker sort of way.

"We need to talk privately," said Wheeler. "Let me buy you a shoeshine." They'd walked down Connecticut Avenue from the trade association's headquarters to Washington's stately Mayflower Hotel, the site of many presidential inaugural balls and a fine shoeshine stand. Sitting side by side in the stand's tall chairs, they were a most unlikely looking Washington power duo: one a clean-shaven, bespectacled power-lobbyist wearing a finely tailored dark suit, white custom-made shirt with "TEW" monogrammed above the pocket, shiny cufflinks clasping white French cuffs; the other a gray-bearded, shaggy-haired epidemiologist, in a tweedy sport jacket, shirt, and sweater.

"You and I are tied at the hip on this," Wheeler said, speaking candidly as if oblivious to the presence of the two middle-aged gentlemen who were in front of them, shining their loafers. "If you succeed, I succeed. If you don't succeed, I don't succeed."

What the lobbyist wanted, Carlo believed, was not Siamese-twin comradery, but political separation: an ample degree of detachment from the bad news and himself. When the new scientific findings had to be reported to his powerful multibillion-dollar association's board members, Wheeler told Carlo to deliver the news—at a board meeting that was closed to the public and the press.

And to that very public worldwide scientific colloquium in Long Beach, Wheeler sent CTIA vice president Jo-Anne Basile to represent the industry and do what she could to stroke the scientists and spin the journalists. Meanwhile, Wheeler remained at his Washington, D.C. command headquarters 3,000 miles away—where, being a Civil War buff, he was putting the finishing touches on a book, to be published by the end of 1999, about leadership lessons that 21st-century business executives can learn from the generals of the Union and Confederate armies.

Wheeler's new war-game strategy for the cell phone industry seemed clear: regroup, retrench—but never retreat.

So it was that in Long Beach, when Carlo finished issuing his public call for new industry safety standards, all eyes seemed to shift from the podium to Wheeler's CTIA representative in the audience. Jo-Anne Basile rose from her seat; surely she was expected to say something in response. "You have caused us a few sleepless nights," she told Carlo, who was still standing at the podium. Her public response in that hall indeed emphasized the civil, rather than the war.

But shortly after the meeting adjourned, Basile and Carlo came face-to-face in the corridor—and their chance meeting quickly erupted into a shouting match. The dark-haired industry rep blasted Carlo for daring to make a public call for a new radiation-emission standard without first clearing it with the CTIA.

Carlo responded that the industry was failing to meet its public-health responsibility—"and that's shameful."

Basile fired back: "How dare you talk to us like that after all the money you've been paid!"

Suddenly, Carlo was aware that people in the corridor had stopped to listen. The fact that his research project was funded by the industry had always been a sensitive point with Carlo. He saw himself as an independent researcher whose only goal was to follow the scientific evidence and protect public health. Yet he knew that people had long been saying he was industry's hired hand. Over the past five-plus years, he'd heard it every time he issued a public statement that basically brought aid and comfort to the industry CEOs by discounting some of the early scientific scare studies on the grounds that they were, in fact, flawed.

"I take my job seriously," Carlo said, now making sure he was speaking loudly enough for all eavesdroppers to hear. "Money has nothing to do with that."

A SUIT HEARD 'ROUND THE WORLD

It is easy to pinpoint the moment that set in motion the chain of events that caused Wheeler to put Carlo in that job—and eventually resulted in an epic collision of science and politics.

It was January 21, 1993, and Washington was alive with new beginnings. A new president had just been inaugurated. A new Congress had just been installed. At both ends of Pennsylvania Avenue, the powers of the nation's capital were still celebrating their good fortune. But at CTIA headquarters Tom Wheeler, the newly appointed president and chief lobbyist of the powerful trade group, was scrambling his troops in an effort to stave off an industry crisis and, in fact, a nationwide panic.

On this politically charged day after Inauguration Day, CNN talk-show host Larry King wound up making major news by booking a guest who had nothing to do with politics at all—a private citizen from Florida whose story ignited a crisis that would shake the power brokers from Washington to Wall Street. David Reynard of Tampa, Florida, told Larry King why he was filing a lawsuit naming cell phone industry companies as defendants. Reynard was alleging that his late wife, Susan, had suffered a fatal brain tumor due to her repeated use of her cellular telephone.

"Suit Over Cellular Radiation Raises Hazard Questions," said a headline in *The Los Angeles Times*.

"Cellular Phone Safety Concerns Hammer Stocks," said *The Wall Street Journal*. In the week following that *Larry King Live* interview, Motorola's stock prices dropped by $5.37 to $50.50 after a brokerage house lowered the stock rating for the nation's largest cellular phone manufacturer. And stock prices for cellular service provider companies dropped as well: McCaw Cellular stock fell $2.87 to $33, and Fleet Call stock fell $1.62 to $20.52.

Meanwhile, the cellular phone industry had its own headline spin: "CNN Runs Scare Story," the CTIA Newsletter had dismissively declared. But the industry's problem was that the story really did seem scary to millions of cell phone users. News of the lawsuit, and its hard-to-prove claim, quickly became a national and international news sensation. It triggered an instant inquiry from a subcommittee chairman in the U.S. Congress, and it quickly caught the attention of an even more powerful and influential opinion-shaper: Jay Leno made it part of his late-night TV comedy monologue. Before the year would end, Tom Wheeler would write a memo to his top advisers that aptly characterized his beleaguered industry's view of its public enemy: "The Hydra-Headed Cancer Scare."

WHEN CARLO MET WHEELER

It was, in a sense, a fluke that first brought George Carlo and Tom Wheeler together in 1993. But in another sense it was the sort of happenstance that actually occurs just about every day somewhere in the nation's capital. The two were introduced by a public-relations man who was trying to become a power broker between industry and government.

In the spring of 1993, Carlo was at a bed-and-breakfast inn he owned on the Maryland shore of the Chesapeake Bay when he received a telephone call from Mark Shannon, of the Ketchum public relations firm in Washington, D.C., who knew of Carlo's work as a pathologist and epidemiologist willing to get involved in the business of giving advice to industries. Shannon wanted to consult—to get a few expert thoughts and phrases. He was about to meet with the cell phone industry's chief lobbyist, when he would be making a pitch on damage control in the hopes of landing a lucrative PR contract.

Carlo listened, then gave some quick advice. Almost as an afterthought, Shannon asked Carlo to come along to the meeting with Wheeler, thinking this would add scientific credibility to his PR pitch. And so, a few days later, Carlo found himself in an office building on 21st Street NW that at the time housed the CTIA. (As the industry's fortunes soared in the years to come, Wheeler moved the association into its current headquarters on Connecticut Avenue.) A dark-haired, 46-year-old career lobbyist, Wheeler had already earned a reputation for his ability to move within Washington's corridors of power. He'd become known as one of the capital's most savvy movers and shakers when he served for five years as president of the cable television industry's trade association. He'd been with the grocery manufacturers' trade association before that. In short, Wheeler had long ago proven his mastery of the Washington art of political science. In the coming years of crisis in the cell phone industry, he would expand his skills into the selective use of highly political science.

Wheeler struck Carlo as a formidable and commanding presence, a take-charge CEO. Indeed, he was that. His book, entitled *Leadership Lessons from the Civil War*, summed up his own management style in his choice of chapter titles: "Lesson One: Dare to Fail; Don't Confuse Victory With Avoiding a Loss . . . Lesson Three: Yesterday's Tactics Make Today's Defeats; Embrace Change . . . Lesson Five: Information Is Only Critical If It Is Used

Properly; Use It or Lose It . . . Lesson Seven: Small Skirmishes Decide Great Battles; The Power of the Individual . . . Lesson Nine: If You Can't Win . . . Change the Rules; Think Anew."

Wheeler's literary table of contents became his literal battle plan when the cellular telecommunications industry came under attack. The standard Washington response of any political organization being attacked in the media is to mount a strong, well-financed PR campaign. And to be sure, Wheeler would see to it that his forces mounted one of the best. But in January 1993, with his fledgling industry buffeted by the forces of CNN, the Congress, and Leno, Wheeler had come to another quick conclusion: He was not about to be seen just sitting in his headquarters, firing off press-release popguns in response to the salvos of media allegations of this life-or-death magnitude—allegations that there might be a connection between cell phones and brain tumors.

To keep the public buying and the government regulators at bay, Wheeler decided that the CTIA must mount its own initiative quickly and publicly. Just one week after that Larry King interview on CNN, Wheeler held a press conference and announced that the industry would be sponsoring a huge industry-funded research program.

"Despite the many research studies showing that cellular is safe, it has become necessary to reassure those whose doubts have been raised by this scare," Wheeler said in his January 29, 1993, press statement. "It is time for truth and good science to replace emotional videotape and unsupported allegations. Therefore, the cellular communications industry is today announcing that it will fund research to re-validate the findings of the existing studies, which have found that the radio waves from cellular phones are safe."

Reassurance—a scientifically vouchsafed guarantee of cell phone safety—was what the research program was about. The cell phone industry would pay $15 to $25 million over three to five years for a scientific study that would be expected to "re-validate" previous findings that "cellular phones are safe."

Wheeler began hunting for the right person to oversee his research effort. It would be two months before his announcement that he had found that man in a 39-year-old Washington-based epidemiologist, Dr. George Carlo, an adjunct professor of epidemiology at The

George Washington University School of Medicine, who held doctorates in both pathology and law. It would be years before George Carlo would come to understand the message that was implicit in the wording of Wheeler's initial announcement of the research program—that he and Wheeler did not exactly share the same sense of mission and purpose for the research effort that Wheeler's trade group was going to finance and Carlo was going to direct.

Looking back, it is also easy to see that Wheeler may have viewed Carlo in a way far different from the way Carlo viewed himself. Wheeler understandably would have deduced from his background check of Carlo that he was hiring a public-health research coordinator who could be counted upon to be an industry kind of guy. After all, in addition to his teaching at the university, Carlo also ran a company that did public-health risk management studies for some prominent corporate clients who themselves had been caught up in controversies. Carlo had performed breast implant studies for Dow Corning that had concluded there was minimal public-health risk to their products. And he had made dioxin studies for the Chlorine Institute that concluded low levels of this chemical did not endanger public health. Wheeler might well have expected that any work Carlo did for the cell phone industry would produce results that would be equally welcomed by the industry executives whose companies were paying for this research.

• • •

In April 1993, Wheeler told Carlo he wanted him to run the cell phone public-health research effort. The deal was drawn up quickly and the two men shook hands on it. Carlo didn't realize this new assignment would prove far more controversial than anything he had ever done before. In fact, he was both pleased and impressed with his new role.

I sensed that Tom Wheeler was one of Washington's most savvy lobbyists. And there was no doubt that he was a forceful leader. I remember the day in Wheeler's office conference room when we came to terms. Tom leaned back in his chair and said to me, "It's a good idea. But I'm not going to be the fall guy if this goes bad . . ."—and suddenly he sprang forward, jabbed his index finger at my solar plexis, and added—"You are!"

• • •

Carlo's appointment by the CTIA to direct its $25-million scientific study project was greeted with little enthusiasm within two sectors that would be crucial to his efforts.

Among the scientific community's narrow circle of recognized researchers and experts in the field, there was widespread surprise and puzzlement at the choice of a fellow they considered an outsider who lacked their expertise. Carlo was a public-health scientist whose specialty was epidemiology—the study of epidemic diseases and their effects on the population. Carlo had never researched, let alone published, anything about bioelectromagnetics—the core discipline of the cell phone radiation controversy. Scientists inspect each other's credentials in the same way that our grandmothers once inspected chickens at the poultry market: they sniff here and there and then shake their heads. So the scientists frankly didn't expect Carlo could accomplish much of significance in this area that was, after all, their life's work and not his.

Among reporters who cover the telecommunications industry, there was a widespread view that Carlo would be a lackey and shill for the cell phone industry. He was, after all, a handpicked expert who they frankly expected would merely provide a polished scientific patina for the industry's standard, high-gloss "no-problem" refrain.

Carlo was very aware of what the scientific community and the news media had been saying about his appointment.

To ease the concerns of the experts who thought he was lacking in credentials, Carlo created two panels of prominent scientists. First, he formed the Science Advisory Group (SAG), and recruited two top experts to work with him. As he recalls, the key to its success was that he was able to convince two top experts to work with him

I first recruited Dr. Arthur (Bill) Guy, perhaps the dean of all bioelectromagnetics scientists and certainly one of the world's foremost experts in the measurement of radio frequency radiation. He was

Emeritus Professor at the University of Washington in Seattle, and had done work for the cellular industry before. The CTIA had suggested that I approach Dr. Guy about participating, but I did not follow the suggestion until I had the opportunity to vet him myself. That process was easy—a simple literature search yielded his name on virtually every important committee over the past two decades, and what seemed like literally hundreds of publications had him in authorship. My first contact with him was at a meeting in Washington, D.C., at CTIA headquarters. He was in town giving advice to CTIA about the siting of base stations for transmitting cellular telephone calls, and I had arranged to meet with him. I believed it was important for me to reach out to him, out of respect for his stature in the field. After the first five minutes I knew we not only wanted him to be part of the SAG, but we needed him. He was a tall, rugged-looking man in his early 60s, with a disarming and low-key demeanor. He looked like a professor, probably the type that every student wanted to work with. We talked radio wave dosimetry, the concept I had developed for the SAG, and what his role would be—the head of all the dosimetry work. He had many tough questions, and I sweated it out for a while. After all, I knew very little about this science, had done no research in it, and I was about to embark on, and in fact head up, the largest program ever attempted in this field, with no prior experience. So I was worried, at first, about trying to impress the man who many people believed was "the field." But when the conversation drifted to salmon fishing, it was clear we were comfortable with each other. He agreed to sign on.

To round out the SAG I recruited Dr. Ian Munro, a world-renowned toxicologist with whom I had worked on a number of prior projects involving herbicides and drug development. A former high-ranking official in the Canadian Health Ministry, Dr. Munro would handle oversight on all nonhuman research in the program. He was also a fisherman.

So the SAG was in place with Dr. Guy overseeing dosimetry, Dr. Munro toxicology, and me covering epidemiology, public health, and general management. We began daily phone conferences and designed our program. This was April of 1993, and we needed to present our overall program to the U.S. Food and Drug Administration (FDA) on May 22.

Carlo made one other major move to ease the concerns of the scientists—and to impress the politicians. He created a Peer Review Board (PRB) that would be headquartered at the Harvard Center for Risk Analysis, led by the respected Dr. John Graham, of the Harvard University School of Public Health. The peer-review group would be comprised of internationally recognized experts who would examine the findings of studies done by or funded by Carlo's SAG project and also review funding proposals from outside researchers. That clearly pleased the officials at the FDA; their agency was already enmeshed in a major controversy over breast implants and did not relish the prospect of having to be on the front lines of two political wars simultaneously.

Those moves enabled Carlo to ease, at least initially, the concerns and jealousies within the scientific community.

To ease the concerns of the journalists who thought he would be an industry shill, Carlo—well, Carlo frankly did not know what, if anything, he could do quickly. He didn't see how he could win the media's respect, or at least a decent interval of benign silence, until he had time to prove himself by doing his job. Carlo knew, deep down, he was not going to be a shill; but then again, truth be told, he always thought, deep down, that his research project would conclude what the early studies and the cell phone industry had always asserted—that there was no evidence that wireless phones cause cancer. So Carlo just went about his job, hoping his SAG and the Harvard-led peer-review group would provide valuable credentials of respect. He believed that if he conducted himself responsibly, he would be judged responsibly.

THOUSANDS OF STUDIES?

The cellular telephone industry unwittingly created the first impossible task for its new research chief. In the wake of the first wave of scare stories in the news media and panic selling of cell phone stocks on Wall Street, the industry had offered instant reassurance.

On January 26, 1993, a senior Motorola executive told reporters that "thousands of studies" had already shown cellular phones were safe. It was a classic overstatement that all in the industry would regret enormously.

News accounts everywhere began referring to the existence of thousands of studies as if it were, in fact, a fact. Carlo found himself swept into the rushing stream of assuring rhetoric, as he too was quoted on several occasions talking about these thousands of studies. Naturally, news reporters began asking the industry to make those 1,000 studies public. Carlo put his staff to work at the task. A research firm was contracted to conduct a huge Internet database search for the thousands of studies. But thousands of studies were not to be found. Months later, the CTIA staff was still scrambling, to no avail.

On July 13, 1993, the CTIA's director of industry relations, Cilie Collins, wrote an urgent plea to Dr. Om Gandhi, of the University of Utah, one of the pioneer scientists in the field of cellular telephone research. "We need copies of any studies that are pertinent to this issue to be available to the press," Collins wrote. "As you know, one of the main causes of the cancer-scare media coverage was that the industry was unable to produce the 'thousands of studies' that have been conducted on the cellular phone frequency."

There was of course only one reason why the industry was never able to produce evidence of those "thousands of studies" that said mobile phones were safe: The studies did not exist. The entire industry regretted its initial reflexive-response "thousands of studies" posture, as journalists began to view a bit more skeptically every assertion the industry would make during the coming years of political and scientific war-games.

Belated Background Check

In mid-May 1993 Wheeler opened a meeting of his top policy and public-relations advisers and his science adviser, Carlo, in the CTIA boardroom by announcing that their agenda for the session consisted of two items: One was the credibility of Carlo, and the other was the credibility of the SAG program Carlo had been appointed to run just a month earlier. Wheeler had a habit of writing meticulous, printed notes in his day-timer calendar, and he was reading from those notes.

"What do you have to say about the flap in *Science* magazine?" Wheeler was looking directly at Carlo, who was clearly caught off guard. He was referring to a magazine article about a controversy that had caused Carlo to end his six-year relationship with the Chlorine Institute—after the industry's public-relations representatives had put Carlo's name on top of a PR paper that he had not only never written but never even seen. Wheeler had never mentioned this issue before—it seemed obvious to Carlo that someone had brought the matter to Wheeler's attention as a way of questioning whether Carlo should be running the industry's science research program.

Carlo explained the dioxin uproar: In February 1991 *Science* carried a story about a controversy that had erupted after publication of a paper listing Carlo as its author. It characterized the views of a scientific advisory group sponsored by the U.S. Environmental Protection Agency (EPA) official. When one participant wrote an angry letter to Carlo and others protesting these comments Carlo had allegedly written, Carlo was shocked. He had not written the paper, and had not even seen it before. All he had done was write a summary of a conference he had attended which carried the notation that "the meeting reinforced the notion that dioxin is much less toxic to humans than originally believed." That phrase became part of a new chlorine industry position paper The chlorine industry officials told Carlo they had put his name atop the paper as its author to give the document added credibility. The industry and the PR firm each said they thought the other had told Carlo about it—which of course would still have been unacceptable because he simply hadn't written the document.

Wheeler listened intently to Carlo's explanation.

I had the sense that he saw this as an issue on two levels: First, could he trust me to carry his interests forward? It surely appeared to him that I may have turned against the Chlorine Institute. And second, would there be any spillover onto the SAG program? Had the "flap" as

Wheeler termed it, harmed my reputation as a scientist?

I tried to reassure him on both counts. I saw the "flap" as a simple indication that I played by the rules, and that I expected those with whom I was working to play by the rules as well. The feedback I had gotten about the controversy from my scientific colleagues in the months following the Science *article was overwhelmingly positive. Many of my friends in the chemical industry thought the situation was unfortunate but did not blame me for my response.*

"I am satisfied with your explanation, George, but I still don't think you can be out there alone on this," Tom said.

"That is precisely why we'll have a Peer Review Board," I answered. "With some of the world's top scientists helping us, our science will be above reproach. It will speak for itself."

"Not good enough," Tom responded. "Politically, and from a public-relations view, we need more cover."

I disagreed with him and argued that everyone's interests would be served if we trusted in the science and did the best science we could. Everything would flow from that. We didn't need to contaminate this with politics.

Tom gave me an angry rebuke: "You do the science. I'll take care of the politics."

CHAPTER TWO

FOLLOW-THE-SCIENCE: *NO PROBLEMS, BUT NO TOOLS*

WHEN CARLO AND HIS TEAM began their research in 1993, the state of the science about whether cellular phone radiation could produce health effects in humans was downright primitive. Basically, the scientists—and the government regulators—treated cell phones as if they were akin to microwave ovens. The only known way to measure radio wave radiation was to measure the heat that was produced by it—and that led the scientists and regulators to assume that the only way radiation from a cell phone could cause damage would be if it created heat and actually cooked tissue, much the way a microwave oven cooks a pork chop.

At a time when there were already 15 million Americans holding cell phones against their heads each day, science had a simple answer to the question of whether those phones posed a health risk: If the radio frequency radiation from the equipment was not high enough to produce heat that could cook tissue, then the instruments surely posed no health risk. And since no thermal effects had been observed from cellular phones, which operate at low power levels, the scientists and government officials in 1993 were operating from this early consensus bottom line: No heating means no problems.

To understand the risk of cancer associated with cellular telephones, George Carlo, Bill Guy, and Ian Munro—the Science Advisory Group (SAG)—set out to understand the effects of radio waves on DNA, chromosomes, and the human body's ability to repair or adapt to genetic damage when it does occur. They began by designing a research agenda that would examine the effect of radio waves on live human blood cells in test tubes and petri dishes (known as *in vitro* experiments)—tests that could be done in a short time period and would provide a quick indication of dangers that might be inherent in cell phone radiation. The group also focused on the need for experiments that exposed to radio waves the heads of live laboratory animals, mainly rats (known as *in vivo* experiments)—tests that might take more time but which could also be crucial to understanding whether radio waves could indeed cause tumors or other biological changes in living animals, and ultimately in people.

But when the three researchers set about reviewing the limited work that had been done in the field, in order to design their experimental programs, they discovered to their surprise that there was no established set of procedures or devices for doing the precise sort of cell phone exposure experiments that were needed.

Past experiments that had involved human cells in test tubes and petri dishes had no set method for beaming the plume of radiation from a cell phone antenna into the contents of these containers. Nor had any determination been made as to how the radiation actually acted once it penetrated the test tubes and petri dishes—whether it dispersed evenly to reach the blood or tissue inside, and whether it caused that tissue to heat.

Past experiments that involved laboratory rats or mice had not used equipment that was specifically designed to replicate, on these laboratory animals, the manner in which radiation from cell phone antennas reaches, and perhaps penetrates, the human head. Indeed, a number of the experiments did not ascertain whether the radiation actually reached only the animals' heads or perhaps was spread over the animals' entire bodies.

For both the *in vitro* and *in vivo* experiments, there simply were no tools. If Carlo and his group were going to commission a research program that would be truly relevant to investigating the potential problems posed by this new instrument—the cellular telephone—then the three investigators would have to first invent the tools that would assure the research would be valid. It would take more time than had been anticipated and cost more money than had been promised by the industry. But there simply was no other way to do the science correctly.

Looking back, it should not have been surprising that the scientific world had not developed the tools necessary to assess the impact of radio waves on human heads. After all, there seemed little need to develop these tools—when, in fact, the best minds in the field had not even identified that a potential problem existed with the exposure mechanisms and methods used in the early experiments.

NO HEATING, NO PROBLEMS

The general consensus among scientists and standard-setting federal agencies was that if there was inadequate energy to cause heating of biological tissue or breakage of DNA, then there should be little to worry about with respect to human health effects. The best scientific minds in the federal government believed that the only radio frequency radiation waves powerful enough to heat biological tissue occur in microwave ovens—with 100 or more watts of power that propel waves at 2450 megahertz. That was considered the only way that radio wave radiation could cause serious health effects such as cancer.

Scientists did know the chain of thermal and biological events that could lead from the heating of tissue to the development of cancer: When tissue is heated, cellular physiology is altered and there is a breakdown in the basic processes that serve to provide nutrients to cells and control cellular functions. When these functions break down a host of problems can ensue, including mutations and damage to the immune system. Significant heating causes death of cells and severe damage to biological tissue that can impair whole organs, such as the kidneys and liver. Breakage of DNA, which can occur when it's heated, is a potentially serious problem as well; but since all mammals, including humans, have sophisticated systems that repair DNA, the breakage of DNA does not inevitably lead to disease. However, if the rate of DNA breakage is greater than the ability of the body to repair the broken DNA, serious mutations and chromosomal anomalies can occur. And that can lead to the development of immune system problems, birth defects, and cancer.

In 1993 there were a few scientists who had hypothesized that nonthermal levels of radio frequency radiation could theoretically produce biological effects. But this had not been observed in laboratories. And since there was no scientific proof that health effects could be caused by radio waves of low power that produced no significant heat, the government simply did not regulate instruments, such as cell phones, that operated at these lower levels.

The federal government long ago issued rules to assure the safety of microwave ovens—FDA regulations required that the ovens be properly shielded to keep microwaves from leaking outside the appliance. But the only government rules that could be applied to the safety of mobile phones had to do with the measurement of the instrument's known radio wave radiation, which was calculated as an index known as its specific absorption rate, or SAR. The SAR is a complex measurement of how much radiation passes through tissue during a specified time period. And since the calculation of SARs would prove key to their study of cell phone radiation, Carlo, a public-health scientist by training, made it his business to learn all he could about SAR measurement from his inside expert, Bill Guy. But, it was not a subject that lent itself to quick study.

"So the SAR is a measure of heating?" It was April 1993. Bill Guy and I were in the conference room of my townhouse office on N Street.

Dr. Guy patiently tried to explain again. "No, heat is a part of the formula to calculate it, but it measures the amount of energy passing through tissue during a time period. It's more than heating."

"But if it depends on heat, it has to be a measure of heating."

"It could be, but not always," he replied, his professorial demeanor always intact.

"I still don't get it."

"You will. Some day." He laughed, and that signaled the end of my first lesson in the complex field of radio frequency dosimetry.

A PRIMER: HOW RADIATION IS MEASURED

The calculation of the SAR begins by measuring the amount of heat that is generated by radio waves passing through the target tissue—brain cells, for example—at one specified moment. As of 1993, no scientist had observed and confirmed that there were any thermal effects of radio waves at SAR levels below about 40 watts per kilogram (W/kg). And cellular phones operated at SARs far below that level—with power of less than .6 of 1 watt, which yielded SARs of less than 2 W/kg.

That was the rationale—widely accepted, yet scientifically unconfirmed—that led the most influential standard-setting group for radio frequency radiation (the IEEE/American National Standards Institute) to exempt handheld wireless devices such as cellular phones from their recommended guidelines for exposure. This was the same rationale that FDA scientists gave when Carlo asked them why they had no regulations regarding these phones. That was the rationale the government used to explain why they had not required premarket testing before millions of Americans would be holding mobile phones with radiation-emitting antennas against their heads.

This science of the dosimetry of radio waves (how the radio frequency radiation travels into and through people's cells, tissues, and organs) turned out to be a tough study—even for the scientists who made it their life's work. The only easy thing about this science was understanding why it was so tough: Radio waves could not be seen. The only obvious effect of these microwaves was their capacity, with enough power pushing them, to cook food. The more subtle effects of the radio waves could not be seen or identified without very sophisticated tools. And even then many of the effects were only theoretical.

In 1993 the fact that effects of radio waves could not be seen gave a very deceptive sense of reassurance to policy makers, industry leaders, and scientists. Their thinking went something like this: If there was a problem with cellular phones, it would only be a small one—because, after all, we haven't seen anything at all conclusive, even when we look very hard. In that highly speculative climate, it was easy for scientists to take sides. Some scientists believed there were problems with radio waves, other scientists believed there were none. The scientific literature in the early 1990s was loaded with radio frequency (RF) scientists making strident attacks on one another about interpretations of scientific studies and the public-health conclusions that could be drawn from them. It made for good science theater; but it made for confused and even contradictory public policy. And it led to mistakes.

Shattering Test-Tube and Petri-Dish Findings

From the beginning in 1993, Ian Munro, heading up Carlo's Toxicology Working Group, had been concerned that none of the established standardized tests for genetic damage had been developed with radio waves in mind. They had originally been designed for testing genetic damage caused by chemicals, such as drugs, pharmaceuticals, and pollutants. Calculating dosage of these chemicals, which can be found in the blood, while not always easy, can at least be accomplished using conventional measuring devices. But calculating the dosage of radio waves that cells or tissues may receive is a very different matter. A radio wave just passes through tissue; it does not remain there. It can't be measured after the fact; there is no way to confirm how much exposure was really received in a part of the body. Thus, good scientific minds would come up with very different theories—a fact that made for good debates but not always productive and illuminating research.

In the biology experiments, there was no standard system for determining whether cells had been uniformly exposed within a test tube or petri dish. And since scientists used a variety of adapted methods, there was no valid way to compare the results of different studies with one another.

• • •

Early in our research, Bill Guy explained to me that there were many problems with this in vitro *dosimetry, and that I had to keep my expectations within range.*

"George, we are not going to be able to solve all of these problems in the five-year time frame we have," he had told me on more than one occasion.

"I understand that, Bill, but what problems can *we solve?" Being responsible for the entire program, I was continually evaluating and re-evaluating where we would put our resources.*

"We need to solve the big problems. Develop the best exposure systems we can in our time frame, then move on with studies that we can rely on." As *usual, Dr. Guy made things seem simple. All we needed to do was solve the big problems, and the rest would take care of itself.*

The two big problems Bill and his Dosimetry Working Group had at the top of their list were these: We needed a system that would give us a uniform radio wave field that allowed us to actually measure heating that might occur during radio wave exposure. That way, the effect of heating could be controlled or eliminated. And that was essential, because we needed to be certain that any effects we detected were due to the radio waves and not to the generation of heat that might build up inside containers such as test tubes or petri dishes.

We also needed to be able to actually measure how the radio waves behaved in petri dishes, flasks, and test tubes so that our results would not be subject to the theoretical criticisms about uneven wave fields that had plagued the existing science. I agreed. Bill moved forward with a plan to address the big problems.

For starters, Bill Guy convened a two-day meeting in Los Angeles and invited 22 of the world's top experts in radio frequency dosimetry to reach a consensus on developing new exposure systems. Most of them came because Dr. Guy had asked them to attend: Dr. Allen Taflove, famous for his work in developing the Stealth bomber; Dr. Stephen Cleary, whose early work showing proliferation of cells exposed to microwaves propelled the mobile phone/brain cancer question into the public arena; Dr. Camelia Gabriel, coming all the way from Great Britain, regarded as perhaps the world's top expert in how radio waves move through biological tissue; Dr. Henry Lai of the University of Washington, a former student of Dr. Guy's who had done early studies on microwave radiation; Dr. Martin Meltz, a top consultant to the U.S. Air Force on radio wave issues; and Dr. Om Gandhi of the University of Utah, who had pioneered the use of a sophisticated computer modeling technique to measure radio wave energy moving through the human head.

• • •

One of the most valuable insights we gained came from one of our most expensive acquisitions: a new, superfast computer known as the Cray computer, which cost nearly $200,000. It was Dr. C. K. Chou's team at The City of Hope National Medical Center in Los Angeles that first found the computer and identified how we could all put it to good use. And in a matter of months, that computer would provide a crucial breakthrough—one that cast doubt on many earlier studies.

Dr. Guy focused on the petri-dish question, while Dr. Chou's team did the time-intensive modeling of radio wave penetration in various body tissues. As they were working on the exposure systems, Dr. Guy worked from his home in Seattle and had remote access to the Cray computer in Dr. Chou's Los Angeles facility.

After hundreds of hours of runs on the Cray, Dr. Guy made a startling discovery—he perfected a way to actually quantify how the radio waves distributed energy in the petri dishes and test tubes, and what he found was unexpected, to say the least. He flew to Washington, D.C. to present his new findings to us.

Dr. Guy had modeled the distribution of the radiation that would come from a cellular phone antenna, measured as the SAR, in test tubes and petri dishes. With the powerful Cray computer, he had been able to look at all aspects of the energy deposition in those vessels—from the sides, from the top, from the bottom, and at different depths within the vessels. For the petri dishes, either glass or plastic round plates about 4 inches across, with sides about .5 inch high, he took "slices" about every eighth of an inch, moving up the side. For test tubes, he did the same thing, moving up and across the tubes.

He presented his findings by showing us brightly colored slides. We could see that along the sides of the vessels were bright red areas—"hot spots." marking the sites with the highest intensity of radiation. In the center were blue areas—"cold spots," indicating where there was little or no radiation.

"For both the petri dishes and the test tubes, the variation in the amount of radiation in different parts of the vessels is large," Dr. Guy *said. "But there is a pattern. There were hot spots, cold spots, and everything in between."*

He went on to say that these studies showed us two very important things: "First, if we are going to use a petri dish, the only places where we can count on a uniform radiation pattern are at the bottom and at the top of the dish. The radiation patterns in between are all over the map, and it would be impossible to quantify how much radiation living cells floating in the petri dish would receive. Second, in test tubes, the only reliable uniformity is at the bottom of the tube—the bottom .5 to .75 inch. The radiation patterns through the rest of test tube are also all over the map, and would similarly be impossible to quantify."

Nearly a third of the entire center of the petri dish was "cold." The same was true for the test tubes—the center of the tube, where most of the cells or blood would be suspended, had almost no radiation.

Dr. Munro *added: "Cells in suspension would naturally move toward the center of the petri dishes and test tubes. Friction would cause the cells to migrate away from the hard surfaces of the sides of the dishes and tubes."*

Because we were focusing on how to do these studies in the future, we needed these data to help us find a pathway to our studies. Our experts had it figured out in short order: For petri dishes, we would use a layer of cells so they were in the bottom of the dish; for test tubes, we would use only the bottom of the tube.

But there was more.

I saw an even greater implication in what Dr. Guy had uncovered. It had to do with the studies that were already in the literature, and with the ones that had found no health risk from wireless technology. My job was to continually look at the new scientific findings from the perspective of what they meant for people using cellular phones. Most importantly, I had to continually assess whether these new data showed a need for some type of consumer-protection step. The recurring question for me was: On the basis of the new information, is there something that we need to do to protect people? By taking this responsibility for addressing the larger public-health picture, I freed the others on our team from the politics that I knew would follow once we made our findings public. They could do their jobs focusing only on the science in our program, and how to do it the best way possible. Given the stakes—the health of millions of people—we needed every bit of their attention focused in that direction.

Clearly, the politics of this new analysis would be powerful inside the scientific community, the cell phone industry, and the government. The studies that were already in the scientific literature, on which policy makers and scientists alike had continued to rely for reassurance that radio frequency radiation did not cause genetic damage, could be flawed. These studies were integrated into the exposure guideline that the Federal Communications Commission (FCC) had recently put into place regarding mobile phone radiation. Further, these studies were cited by the FDA as weighing against there being adverse health effects from these phones.

After digesting the new data, I re-read the review of the genotoxicity studies in our 1994 Research Agenda and the review of the genetic effect literature that Dr. David Brusick, one of the world's leading genetic toxicology experts, had done for us at the end of 1995. The accumulated data in existence at the time showed the following:

Of the more than 105 studies we had originally identified in the literature that addressed the question of radio wave genetic damage, in only eight were the frequencies, powers, and SARs common to cellular phones. Seven of those eight studies showed no evidence of genetic damage. Six of those seven negative studies were in vitro *studies, funded by the U.S. Air Force and performed by Dr. Marty Meltz of the University of Texas. In those studies, Dr. Meltz used cultured cells that were exposed in either petri dishes or test tubes.*

Given what Dr. Guy had now discovered, it was entirely possible that Dr. Meltz's in vitro *studies that showed no effect were false negatives—that is, they showed no effect because the cells in culture did not actually receive radiation exposure; they could very well have been in the "cold spots" that Dr. Guy had discovered. It was too late now to figure out if that was the case—Dr. Meltz's studies took place in 1987 and 1990—but it certainly raised questions about how much they could be relied upon to attest to the safety of wireless phones.*

Furthermore, if you took the Meltz studies out of the mix that was contributing to the weight of the scientific evidence, then there would be one positive study showing genetic damage in human lymphocytes, and one very old (1982) study of mice showing no genetic damage when the entire animal was exposed to microwaves. The mouse study was probably marginally relevant to wireless phones because at the time it was done cellular phone health effects were not even on the radar screen, and no accommodations for relating the study to cellular phones would have been made by the authors.

For all intents and purposes, we were back to square one with the genetic damage question: We really didn't know anything. Several studies had been published that appeared to have some bearing on the question of genetic effects from cellular phones, but when we scratched beneath the surface we found very little we could rely upon. We had no definitive evidence of genetic damage coming from the use of wireless phones, but we had no reassuring data either.

STEALTH RATS

The most perplexing problem in the radio wave exposure of rats was solved with the help of a most unlikely expert: Dr. Allen

Taflove, of Northwestern University, who achieved scientific and patriotic fame for his pioneering work in developing the radar-evading technology for the U.S. Air Force Stealth bomber.

From the beginning of the program, there was one thing that really bothered me about the microwave exposure experiments on rats. No matter how many of the very knowledgeable, very experienced experts in electromagnetic radiation I talked to, no one really had a convincing answer for what seemed to me to be a basic problem in all of the experiments with regard to cellular phones: In order for us to do toxicology experiments that would provide truly meaningful data, we had to have a system that would expose the heads of animals to radio waves in a way that corresponded to the exposure that penetrates the heads of people when they hold cellular telephones to their ears. We needed to find a system, using microwave frequency and low power, that would expose the animals' smaller skulls in a way that corresponds to the radiation routinely penetrating the larger skulls of humans.

Every time I discussed this with the most knowledgeable people in the field, they replied that there simply was no such system in existence. Clearly, it was up to us to develop a system that would make our science valid.

Scientifically, it would be critical because people were being exposed to what is termed the "near field" of the radiation coming from the antenna of the cellular telephone—within inches of the antenna. In 1993 the only systems that were available for studies of animals exposed the entire animal to radiation. Thus, the exposure was to the "far field," far away from the antenna. But, cellular telephone exposure occurs only in the near field, close to the antenna.

In those early experiments, rats were allowed to run free in waveguides—elongated mesh cages—so the whole rodent body received some level of radiation. Yet, the characteristics of the "near" and "far" fields are very different in both the intensity and the pattern of the radiation plume. We knew that both could be important to what type of effect the radiation would have on people. We also knew that radiation exposures to other parts of a rat's body would alter the way its body functioned. It was possible, therefore, that whole-body exposure was changing the animal's physiology so that the observations being made about cancer development and other health effects were not necessarily what would follow from a person using a cellular telephone, where the radiation exposure is mostly to the head area of the human body. The tools we had were simply not good enough.

In addition, to meet the expectations that we had raised in the minds of the government officials by our presentation to the FDA in May 1993 it would be necessary to make up for the fact that there never had been premarket testing of cell phones. First, experiments had to be done with exposures that were similar to what people using cellular phones would actually experience; second, the studies needed to be able to assess dose–response—whether more of the exposure caused more of the effect. Further, with these two criteria met, the database on cellular phone safety would be similar to what would have existed had there been premarket testing of the phones. I saw that as an important goal for the industry—catching up to where they should have been with the science. This was a priority of our research program.

The task, however, was not so easy. The general scientific consensus was that it would be impossible to scale the radiation plume from a cellular phone down to a level where scientists could replicate in animals what actually happened in people using cellular phones. Scientists believed that with the size difference between the head of a rat and the head of a human, the radiation dose would be far too great in rats to avoid heating of tissue in any scaled-down antenna that could be designed. In fact, during the development of the research agenda, Ron Peterson, a senior level scientist and engineer at Bell Labs who was a member of our Dosimetry Working Group, told me on several occasions that the notion of trying to develop a "head-only exposure system" for rats was foolish. I was personally ridiculed during a presentation I made in late 1994 at the Electromagnetic Energy Alliance meeting because of my insistence that we could *develop a head-only exposure system. A senior U. S. Energy Department scientist commented that I had no experience with radio wave health research, and didn't understand how hard it would be to do what I was seeking. He added that perhaps when I had "grown" in my understanding of his field, I would see it his way.*

Not everyone in the industry shared our concern about the importance of developing a new exposure system. Even Motorola officials, in their attempts to be first with results for all sorts of cellular telephone safety research, were satisfied with a system they had rushed to build in which the radio waves penetrated the head of the rat first, before reaching the rest of the animal's body. With the help of Dr. Neils Kuster in Switzerland and Dr. Ross Adey in California, they developed a system in which an antenna was placed in the center of a number of tubes that held rats in a circular pattern, thus exposing those rats headfirst. There was measurable radiation in the heads of the rats, but the whole body was exposed and the "hot spot"—the place with the highest-peak radiation—was not in the head but further down the back of the rat. At the time, they believed this system to be the best that could be developed.

The Motorola-funded studies based upon that exposure system were cited around the world as proof of the safety of mobile phones. However, with the exposure system they used, the exposures to the heads of the rats were much lower than exposures to the heads of humans using cellular telephones.

There were two problems with the Motorola design: (1) The amount of radiation that reached the head of the rat was measured at only .8 to 1 W/kg—far lower than the radiation level that reaches a human head from a cell phone antenna; and (2) the whole body of the rat received exposure—as the radio waves moved around the rat's body from head to tail, much the way airstreams move around the body of an airplane or automobile in a wind tunnel. The published head exposures from that system were below 1 W/kg SAR, and the average was in the range of .8 W/kg. The SAR guideline published by the FCC is 1.6 W/kg. The bottom line: Studies using this system gave no indication of what type of effects might result at or above the standard. Thus, assurances of safety from these studies were therefore unwarranted until SAR levels closer to the standard had been evaluated and shown to lead to the same types of negative results, if indeed they did.

• • •

As we were working through the theoretical basis for a head-only exposure system, Dr. Guy pointed out that the problem we faced was very similar to the problem the military faced when it developed the Stealth bomber. Radar-absorbing material—the material that prevents radar from bouncing off the airplane, keeping it undetectable to radar—is not dense enough to be a material from which an entire airplane can be made. The friction of flying would cause the airplane to fall apart if it were made only of radar-absorbing material. It was therefore necessary to identify the strategic places on the Stealth bomber where radar-absorbing material would be placed to make the airplane invisible. To do that, it was necessary to have a keen understanding of how radar waves behaved when they hit the airplane.

The person who led the way to the military's understanding of that process was Northwestern University's Dr. Allen Taflove. Dr. Guy reasoned that if Dr. Taflove could figure out where the radar bounced when it hit the airplane, he should also be able to figure out the reverse process—that is, to get a specific bounce of a radio wave, what kind of signal would have to be sent? If you knew the type of signal pattern that you wanted to send, then you could develop an antenna to send that type of signal.

Dr. Guy contacted Dr. Taflove and soon the SAG was using Stealth technology in developing the exposure system. With the help of Dr. Taflove, two antennas—one for cellular phone signals in the 800-megahertz range, and one for 1900-megahertz phones—were developed that allowed head-only exposure of laboratory rats.

However, to use these antennas effectively in experiments it would be necessary for the rats to remain still during exposures that could last for up to two hours. The system that had been developed by Motorola used tubes to hold the rats motionless. Stress can cause the rates of tumors to increase in rats; but Motorola's data made it clear that if the rats were trained from birth to be in the tubes, they were not stressed by the containment. We adopted the Motorola tube approach as the method of restraining the rats during exposures.

• • •

In the politics of science, as in everyday life, the decisions that eventually prove to be the smartest and finest sometimes start out by bringing only grief. So it was with the initial decisions of Carlo and his top advisers, Drs. Guy and Munro, to develop new research tools. When the SAG did not rush immediately to commission costly new research projects, critics were quick to criticize and harsh in what they had to say. And George Carlo bore the brunt of all their ire. The leading trade press—*Microwave News* and *Radio Communications Reports*—carried criticisms that painted him as industry's hired hand who was deliberately stalling for time, and questioned what was happening with the money that had been earmarked for research. The FDA looked skeptically at the industry-financed effort and so did those few in Congress who cared enough to look.

I had already been criticized by the trade press and some industry people for moving too slowly to put studies in the field. As I realized the extent of the delay that would result from our taking this prudent and scientifically sound approach of developing the first properly designed tools, I knew that it would be at least 18 months before we would actually have experiments underway. I knew the criticism in the media would be intense. But I was convinced that scientifically it was the right thing to do.

Before we were through, we spent more than $8 million developing the in vitro *and* in vivo *exposure system tools that our experts told us* *were necessary. I know that they gave us the tools—tools that have enabled all scientists everywhere to get the best possible analysis of the impact of wireless technology on DNA and on human health.*

CHAPTER THREE

FOLLOW-THE-POLITICS: *HIGH AND INSIDE*

THE DOOR OUTSIDE the lobby at the CTIA's 21st Street offices was locked; it was Sunday morning, a little before 10:00 A.M. on July 11, 1993. Carlo went to a pay phone in the open courtyard of the big square office building and called the number that Tom Wheeler's deputy had given him. Soon someone appeared to let Carlo in. By the time he walked into Tom Wheeler's conference room, Carlo realized he was the last to arrive.

Two of Washington's most famous media spinners—Jody Powell and Ron Nessen—were already there, along with Wheeler. Everyone was wearing sport shirts. A coffee urn sat next to Styrofoam cups. The doughnuts had been pretty well picked over. Carlo shook hands all around and took his seat at the long polished wooden conference table. The CTIA had scheduled a press conference for the next Tuesday to report on the research effort to date. They were there on this Sunday morning to decide—and script—what would be said and who would say it.

Carlo had watched Powell and Nessen on TV news back when they were press secretaries for Jimmy Carter and Gerald Ford, respectively. Powell and Nessen had battled each other, sound bite for sound bite, back in 1976 when their bosses were running for president. Now they were sitting together talking like the best of friends, plotting with, not against, each other, to win the next campaign for their new boss, President Wheeler. Powell was the head of Powell Tate, one of Washington's hottest powerhouse PR firms; Nessen was the CTIA's vice president for communications.

I knew Ron had to be there on a Sunday morning, but Jody's presence meant that this cellular phone issue was bigger than I thought. I was impressed. Nessen, Powell, Wheeler—I was blown away to be sitting there at the table with those guys. I felt like a Washington rookie in his first game in the big leagues.

For the next half-hour they reviewed the plan for the staging of the press conference. Carlo was told where he was to stand while Wheeler was giving his opening—offstage to his right. After Carlo was introduced he would walk to the podium, join Wheeler, and give his remarks. There would be a blue background curtain. CTIA's logo would be prominently placed. There would be time for questions and answers, but they'd be limited—"We don't want this thing to get away from us," someone remarked. Following the press conference, there would be satellite hookups to facilitate interviews with local TV anchors around the country. The event would be videotaped in its entirety so all CTIA member companies would have copies.

No question, this was going to be a Wheeler show. I was going to be a prop.

"George, what can you say about the work you have done so far?" Ron asked me.

"So far, so good," I replied. "We haven't seen anything in the data to suggest a problem."

"What is that based on?" Jody asked.

"We have reviewed about 400 papers, and there are no 'red flags.' And we are still reviewing more."

Wheeler shook his head. He thought it was too wishy-washy for me to say "So far, no problems." He said it leaves open the possibility that there may be problems. "We need to say phones are safe," he said. "We are here to reassure our customers."

I didn't see it that way. My job was to follow the science. If the public was reassured by that—and they should be—then good.

But after Wheeler spoke the rest of the group chimed in, one by one, with a critique of my every word. "So far, so good" is not strong enough. "Haven't seen anything yet" sounds alarming. I shouldn't say we haven't found any "red flags"—it's not a reassuring image.

They spent the better part of an hour trying to fine-tune my words.

I wanted them to understand my belief that we had to convey to the public what we had, not something we didn't have.

*"Remember that our whole program is aimed at looking for problems. We cannot prove the safety of cellular phones—*ever.*" I repeated over and over. "We can never say phones are safe. The most reassuring thing we can say is that we are looking very hard for problems and so far we haven't found any. But we have to keep looking."*

Carlo tried to explain the situation as he saw it. Cellular phones had not been tested for safety before they went onto the market. By the summer of 1993 there were more than 15 million people in the United States using cell phones that could not be guaranteed safe. Either the FDA, the FCC, and the industry had failed to do their job in ensuring that consumers were safe, or there was a major public perception problem about dangers that were not there—and that had to be addressed.

"I believe the science tends to tip the balance in favor of this being a perception problem," Carlo had said. Indeed, the scientific data so far indicated that the emissions from cell phones were so low that they could cause no health changes in humans or even lab animals. "But we still need to be completely accurate. We have to try to convey to people that we will do all we can to find any health problems early enough to limit their potential damage."

Wheeler and his PR people had not intended this meeting to be a discussion of science. This was supposed to be a rehearsal for a press conference that was scheduled to happen in less than 48 hours. Wheeler listened patiently to what Carlo said. Carlo was not going to budge from his position—"So far, so good" was the only message he would deliver at the press conference.

Carlo was excused from the meeting shortly before 3:00 P.M. A couple of things had been decided by then. First, the press conference was not going to be held on Tuesday; more time was needed to prepare. It was postponed a few days, until Friday, July 16. Second, Carlo's role in the press conference was going to be limited to the purely scientific issues—something he fully supported.

• • •

On Friday, July 16, 1993, the cell phone industry proudly trotted out its science advisory team to give what was billed as an interim report. It was stamped: "Safety Update—Fast Facts: Portable Cell Phone Safety." The report stated that cell phones "fall within the safety standards of the Federal Communications Commission." Left unsaid in the report was the fact that the FCC had also declared that it does not consider itself the "expert agency" for evaluating health effects. Moreover, at that time, mobile phones were exempt from FCC regulations. The report stated that "scientists and government regulators have found no evidence that portable cellular phones cause health problems."

Not leaving it to chance that a reporter might miss the central point, the CTIA document, prepared under the guidance of these major-league spin-masters, carried a huge, bold-typeface conclusion: "**Rest assured. Cellular telephones are safe!**"

Carlo had no qualms about being front and center in that industry press briefing. After all, that's what his evidence to date showed. He was still confident that there were indeed no problems and that his further work over the next three to five years would prove just that.

• • •

Three days later, on July 19, the FDA's top official overseeing the mobile phone matter sent Wheeler a letter that sharply reprimanded the CTIA's president and top lobbyist for what he had said and the way he had said it. Dr. Elizabeth D. Jacobson, deputy director for science at the FDA's Center for Devices and Radiological Health minced no words:

> July 19, 1993
>
> Dear Mr. Wheeler:
>
> I am writing to let you know that we were concerned about two important aspects of your press conference on July 16 concerning the safety of cellular phones, and to ask that you carefully consider the following comments when you make future statements to the press.
>
> First, both the written press statements and your verbal comments during the conference seemed to display an unwarranted confidence that these products will be found to be absolutely safe. In fact, the unremittingly upbeat tone of the press packet strongly implies that there can be no hazard, leading the reader to wonder why any further research would be needed at all. (Some readers might also wonder how impartial the research can be when its stated goal is "a determination to reassure consumers," and when the research sponsors predict in advance that "we expect the new research to reach the same conclusion, that cellular phones are safe.")
>
> More specifically, your press packet selectively quotes from our Talk Paper of February 4 in order to imply that FDA believes that cellular phones are "safe" ("There is no proof at this point that cellular phones can be harmful"). In fact, the same Talk Paper also states, "There is not enough evidence to know for sure, either way." Our position, as we have stated it before, is this: Although there is no direct evidence linking cellular phones with harmful effects in humans, a few animal studies suggest that such effects could exist. It is simply too soon to assume that cellular phones are perfectly safe, or that they are haz-

ardous—either assumption would be premature. This is precisely why additional research is needed.

We are even more concerned that your press statements did not accurately characterize the relationship between CTIA and the FDA ("CTIA has asked the Food and Drug Administration to review and validate this new research to ensure its credibility"). It goes without saying that we would review your data and provide comment on it—we view that as part of our job as a regulatory agency. But since it is not yet clear whether we will help to direct the research program, it is premature to state that we will credential the research.

To sum up, Mr. Wheeler, our role as public health agency is to protect health and safety, not to "reassure consumers." I think it is very important that the public understand where we stand in evaluating the possibility that cellular phones might pose a health risk. I am concerned that your July 16 press conference did not serve that end as well as it should have.

Sincerely yours,
Elizabeth D. Jacobson, Ph.D.
Deputy Director for Science
Center for Devices and Radiological Health
Food and Drug Administration
Rockville, MD

• • •

Meanwhile, over at Powell Tate, PR professionals continued to make crisis control their business. On April 4, 1994, Powell Tate's Kathleen Lobb sent Carlo a memo titled: "Materials for Crisis Communications Plan." She wrote: "As you know, Powell Tate is assisting CTIA in developing a crisis communications plan and handbook that it will make available to its members in order to be prepared if and when an industry crisis occurs." They were in crisis management mode at the CTIA in those days when the media was focusing on cell phones and cancer. And at this time, when all industry hands were pitching in, Carlo agreed to write the introduction for the industry's guidebook for company PR officials.

Looking back, I can see now that writing that introduction was a mistake. I considered myself independent—and should have stayed at arm's length from the industry and its PR operation. By doing that, I clearly sent a wrong signal to some—especially to the trade-association journalists who saw the handbook.

• • •

Carlo would soon discover that there were strange adversaries and even stranger allies in this swirl of controversies over how best to navigate the uncharted science of cell phone radiation. For while the government regulators and the industry at times seemed to be rivals, there were adversarial relationships that were perhaps even more intense within the industry. For example, there was friction between Motorola, the leading U.S. manufacturer of cellular phones, and the Cellular Telecommunications Industry Association, its top officials and even its research director—Carlo.

Chuck Eger had a very difficult job at Motorola—he was supposed to keep an eye on me. He explained that to me shortly after we met in 1993. Chuck had been a Motorola attorney stationed in Phoenix, Arizona, but he was re-assigned to the Motorola Washington office and he told me his job was to keep tabs on what I was doing and what I was planning to do. We actually got along quite well. He would come by my office at the end of the workday, on Thursdays or Fridays, and we would end up at The Tabard Inn, a restaurant and bar a few doors down the street from my office on N Street in northwest Washington. We had very frank discussions about the program—and about life in general. Through Chuck, I came to understand a lot about the politics of the industry. Motorola did not trust CTIA because it was primarily the trade association for cellular phones service providers, and Motorola was a manufacturer. To Motorola, I was CTIA.

One night, in late 1993, Chuck confided that Motorola was also concerned about a scientist in California, Dr. Asher Sheppard, who was testifying at a number of public meetings about the potential health risks of cellular phone base stations. Chuck had previously sent me some recent testimony from one of those hearings and had asked me to comment. I told him it appeared to me that Dr. Sheppard was quite knowledgeable, and that his arguments were persuasive. Although I had not yet had any substantive discussions with Sheppard, my view was that he was someone to be reckoned with. Chuck asked me point blank: "What do you think we should do?" My answer was equally direct: "This is a no-brainer—hire him and get him involved." A short time later, Sheppard was signed as a paid Motorola consultant. Then he was nominated by Motorola to serve on our Peer Review Board—an appointment I approved . During the subsequent years in the program, Sheppard made a number of good contributions to our program from his position on the Peer Review Board. But I always viewed him as Motorola's guy on the board.

THE POWER OF THE PURSE STRINGS

Inside the FDA, in 1993, FDA official Mays Swicord was proposing a takeover. He was pushing hard for a different sort of research structure—one that would put him in charge, with the power to control the research program and dole out the research money to the scientists of his choosing throughout the world. All he wanted the cell phone industry to do was to fund the entire research effort, under a special formal agreement called a CRADA, which is government shorthand for Cooperative Research and Development Agreement. Carlo would have functioned as a mere coordinator, not a controller, of the research agenda.

Swicord had been quite critical of the industry's research program and Carlo's management of it. And he made his views known both inside and outside the government.

Inside the government, at a June, 16, 1993, interagency meeting of executive branch organizations that had interest in the cell phone research effort, Swicord put himself in the position of briefing officials from other agencies on his interpretation of what the industry and Carlo were saying. Among those agencies represented were the EPA, the FCC, the FDA, and the National Institute of

Environmental Health Sciences (NIEHS). According to a government summary of the session, "Swicord conveyed to the participants Dr. Carlo's opinion that industry should be in control. Swicord further stated that, after much discussion, a compromise on a dual approval process would be acceptable to Dr. Carlo."

(Actually, Carlo's own view was that his research program was not subject to control *by* the industry; he believed his program could function as a research effort that was genuinely independent of the industry. It was, frankly, the naive view of a public-health scientist unsophisticated in the power ways of Washington politics. It simply did not strike him as inconsistent to believe, and to expect outsiders to believe, that he could run an independent research effort while still being welcomed into the inner circle of the industry's lobbyist decision makers.)

Swicord went on, in that interagency meeting, to present his own preference that an agreement be drawn up by which the industry would provide the money and the government (Swicord) would be in control. The government's summary memo said, "Swicord discussed the proposal to establish a CRADA to bring funds into the Government . . . Dr. Carlo's organization could act as the Secretariat for the IPAG [Interagency Project Advisory Group] . . ."

After the June 16 meeting, Carlo scheduled a major meeting in North Carolina of scientists who were renowned in the field and officials from government agencies who were following the matter. Carlo and CTIA officials believed that Swicord was urging officials from the other government agencies not to attend the meeting because the SAG was too closely tied to the industry. Soon, Carlo watched in dismay as letters came in from the alphabet soup of agencies—FDA, FCC, EPA, NCI (National Cancer Institute), and so on—all expressing regrets that they would be unable to attend the North Carolina conference.

Outside the government, Swicord was a source for a spate of news articles in industry-wide publications that ranged from skeptical to critical of Carlo's industry-funded research effort. The chief concern raised in these news articles was about the very real fact that Carlo had not yet either commissioned or funded scientific research projects. This skepticism, fed by Swicord's government-regulator stature, gained credence among the journalists who covered the industry. It also began to be shared by many in the scientific community who had hoped and expected to be receiving funds from Carlo's effort. Then CNN and *The Wall Street Journal* ran reports that conveyed the same skeptical tone about the lack of progress in the research effort.

Carlo was being out-PRed. It had begun as soon as his appointment to head the CTIA research program was announced; it was a problem that would persist throughout his tenure. Carlo simply was not getting his message across effectively or convincingly to the journalists or scientists. He felt it would be scientifically wrongheaded and wasteful to rush to commission studies when he knew the testing procedures of those new studies would likely be flawed. The results of those studies, no matter how they turned out, would likely be of little value and perhaps even be misleading.

On October 4, 1993, Carlo wrote a memo to Wheeler and other top CTIA officials. He began by focusing on the PR problem—and eventually got to the heart of the money and turf battle that was driving the public-relations battle:

> "After some further thought and perhaps stewing, I would like to suggest that we consider the following as further steps to set the record straight on our 'little bump in the road' . . . Include a frank letter from the SAG chairman [that's Carlo] in the next [CTIA] newsletter . . . 'to clear up the misinformation and rumors' . . . [Write] a short Letter to the Editor of the *WSJ* [*Wall Street Journal*] and CNN clarifying the misinformation regarding the Blue Ribbon Panel [the Peer Review Board], role of FDA and relationship with FDA regarding research program . . ."

Then in November, the CTIA's Wheeler got word that the CBS News magazine show *Eye to Eye* (which was then anchored by Connie Chung, and which had as its executive producer Andrew Heyward, who would soon become CBS News president) was planning a hard-hitting report about the cell phone controversy. And Wheeler believed much of the CBS information was being fed to the journalists by Swicord. That was too much to take without striking back.

On November 26, 1993, Wheeler wrote a memo to his top associates outlining his own battle plan. Interestingly, this memo did not take the form of most of his other memos, in that he did not mention that he was its author. But it was indeed Wheeler's—and while he started out talking about himself in the third person, using just his initials, "TEW," he wound up referring to himself as "I."

Wheeler's memo revealed a most unusual effort—the industry (CTIA) had teamed up with the government agency that was regulating it (FDA) to foil the media's effort to inform the public about what was happening. He also came up with an attention-getting title that indicated the industry's collectively beleaguered state of mind: "Dealing With Hydra-Headed Cancer Scare."

Wheeler began his memo by discussing what to do about CBS News' "Eye to Eye" show. "Until they [CBS News] indicate that they are looking for something more than a story of a fight between FDA & CTIA, we won't appear on camera," Wheeler wrote. And he added that "the FDA will do the same." Wheeler wrote that he had information that CBS journalists had three letters from FDA that had been written to "Carlo/Wheeler" and he said that these letters had been given to the journalists by an FDA science official, Mays Swicord. Wheeler said he would ask the FDA's Dr. Elizabeth Jacobson to write him another letter that would make it clear that, while there was "no secret about bumps in the road" between the agency and the industry, that things are "back on even keel" and that the FDA will be participating in the industry's planned meeting that December.

Wheeler's memo went on to discuss his strategy for dealing with one of the central issues CBS was apparently exploring, based on the questions the journalists had already been asking the industry. Namely: whether the research should be controlled by the government, with the industry just supplying the funding, as the FDA's Swicord wanted. Back on October 4, Carlo had outlined a bluntly-worded position in his memo to Wheeler on why the industry and FDA should never get into a situation where the industry would be giving funds directly to the government regulators to fund research:

"I suggest that the following points be made whenever we have the chance (perhaps these should be cleaned up and sent to member companies as well as a basis for responses to inquiries) . . . It is inappropriate and unprecedented for an industry who could be regulated by FDA to give FDA money for research—such a practice smacks of impropriety on its face; if the FDA decides not to regulate, the headlines could read, 'Industry Buys Off FDA—No Regs on Cellular Phones'; no one wins in such a scenario; the industry never considered such an arrangement, and it was *never* "on the table" as was reported in *Microwave News*; the FDA is not a monolithic group, and those in charge are comfortable with the industry research approach; propagation of the industry-funded FDA research concept is a self-serving ploy espoused by one or two midlevel bureaucrats within the [federal] agency who would benefit by the receipt of external research funds. . . ."

Carlo was pleased that his position became the position Wheeler went on to take in his November 26 memo on "Dealing with the Hydra-Headed Cancer Scare" He wrote that CBS had raised questions about why the industry isn't just giving money to the FDA to do its research. "The reason CTIA hasn't given money to FDA (as Mays wants) is because it would be inappropriate to do so," Wheeler wrote. He added: "Jacobson agrees that they can't be research collaborator and regulator simultaneously. Wheeler added that the CTIA had always said that it would fund its own research program and that the trade association called upon the federal officials to appoint their own "blue ribbon panel" to review the industry's research and attest to its objectivity.

Wheeler went on, in his memo to issue a battle-planning command that would have made a Civil War general proud. He said he wanted Carlo, and perhaps the peer reviewer officials at Harvard University to write a memo for him that would be an "integrated assessment as if they were the FDA"—in other words, he wanted a war-gaming memo. "In other words, I want to know NOW where we will be vulnerable so that we can attempt to mitigate that vulnerability now," Wheeler wrote. He made clear he was trying to anticipate and then out-maneuver the FDA official and then-industry nemesis, Mays Swicord. "Let's face it, Mays can't just rollover [sic], he'll have to find something to snipe at," Wheeler wrote. (Despite this, Swicord told Martin Schram, "I did not ever see myself as a thorn in anybody's side.") "Most probably, that sniping will be in the form of some design factor . . . let's anticipate and figure out how to control the opportunity."

Wheeler also pushed his industry team to begin to identify areas for research projects. He added: "I don't want to be sitting around sucking eggs in early January if we know what the next steps should be."

• • •

Seven years later, in 2000, all of that would be topsy-turvy. Wheeler, unhappy with the unexpectedly tough scientific conclusions Carlo's research program would reach about the health risks of cell phones, would end his industry's ties to Carlo—and would reverse his position in which he had made a strong case that it was wrong to have the industry fund the government regulators' research. And Wheeler would rush to have his industry team up with the FDA in a research agreement—just like the idea Mays Swicord had pushed for and Wheeler had called "inappropriate." (As a Civil War general might say, "If you can't win, change the rules.")

But if you think this finally gave Mays Swicord the control he had long sought, well, you'd be wrong. For the cell phone industry found another way to smooth relations with Swicord: It hired him. Motorola offered Swicord a job in its scientific research program. So Swicord left the FDA and, participated in a longstanding custom by passing through Washington's well-known revolving door, into a job with an industry he once helped to regulate.

If you think that in the year 2000 the FDA's new regulators responded by stepping up their independent, adversarial regulatory posture, and that they maintained a healthy skepticism toward the industry they were regulating, you'd be wrong again. Surprisingly, in 2000 (as is discussed in Chapter Sixteen), a new regime of FDA regulators would drop their independent skepticism and begin issuing the sort of no-problem, not-to-worry public assurances to the news media that only the industry had been making back in 1993.

And if you think that the news media properly and quickly pounced on the FDA for its surprising new posture of issuing statements that mirrored those of the industry it was supposed to be regulating, well, you are indeed having a bad day. For most of the news media that is supposed to be the public's ultimate watchdog seemed to have fallen asleep on its watch—and seemed unaware that the FDA regulators in the year 2000 were voicing assurances that had the same tone and tenor as those the CTIA had been issuing in 1993.

• • •

Wheeler ended his November 26, 1993 memo with some final thoughts about Carlo's role. He made it clear that he saw Carlo as an industry insider—and that the FDA had told him it thought of Carlo in much the same way. "One of the things that concerns FDA is that George is an advocate not a scientist," Wheeler wrote. "He's our counsellor on these issues and the man who is managing the process . . . as such he's suspect. . . . It is Harvard which has been retained to manage the research program. George has been retained to advise CTIA and to manage CTIA's component in the overall program."

Wheeler, in his memos of that period, repeatedly sought to capitalize on the lofty name of the Harvard University Center for Risk Analysis, which Carlo had recruited to coordinate the independent peer review process of his research program. Wheeler repeatedly sought to increase the role and involvement of this Harvard tie, knowing it would be perceived as a seal of approval. Carlo felt that any effort to expand the Harvard center's role would inevitably compromise and co-opt its vital independent peer review process. The Harvard center remained at an independent arm's length from the industry's political and scientific agendas.

TRUTH AND LABELING—I

In late 1993 the cell phone industry set up a committee to write a responsible consumer's manual. Its draft document began with a stark, attention-getting heading:

> "IMPORTANT: READ THIS INFORMATION
> BEFORE USING YOUR TRANSPORTABLE
> CELLULAR TELEPHONE."

When the committee duly sent Tom Wheeler a copy of the draft, he fired off a memo expressing his concerns with the wording in several key places in the body of the document—wording in which the industry's own experts seemed to acknowledge, or at least imply, that cell phones could pose health risks. Motorola's vice president of marketing, James P. Caile, quickly responded to Wheeler in a January 13, 1994, memo, sending him the revisions that had been made to address the concerns raised by the lobbyist. (The Motorola executive added that Wheeler was not the only industry insider who'd wanted to change some wording, as he noted: "I know George [Carlo] had indicated some concerns of his own . . .") To his note for Wheeler, Caile attached a copy of the handwritten deletions and insertions that had been made to the original draft. "This represents a significant change from the draft materials on which the labeling committee had previously reached a consensus," Caile concluded.

Indeed, the changes made to meet Wheeler's concerns were significant. Among the deletions were key phrases that had originally been included from scientists—phrases that noted just how mobile phones could potentially cause health risks. In a four-sentence section titled "Efficient Phone Operation," the second and third sentences were scratched out on the draft (to make sure there would be no mistake, in the margin there was a handwritten notation: "Remove"). Here is how the attached draft read, with bold type indicating the second and third sentences that were crossed out, converting it into a simple two-sentence section.

> "Do not operate your transportable cellular telephone when holding the antenna, or when any person is within 4 inches (10 centimeters) of the antenna. **Otherwise you may impair call quality, may cause your phone to operate at a higher power level than is necessary, and may expose that person to RF energy in excess of the levels established by the updated ANSI Standard.**
>
> **"If you want to limit RF exposure even further, you may choose to control the duration of your calls or maintain a distance from the antenna of more than 4 inches (10 centimeters).**
>
> "For best call quality, keep the antenna free from obstructions and point it straight up."

Another section, titled: "Electronic Devices," was originally drafted to address news media reports that emissions from cell phone antennas had been shown to alter the functioning of some heart pacemakers and other medical devices, and even some electronic equipment in cars. The section had been revised with a brief handwritten insertion to make clear the industry's position that any problems which might occur due to emissions from cell phones were the fault of the manufacturers of the medical devices and automobiles (not of the cell phone manufacturers). Here is how the attached draft read, with the handwritten insertion shown in bold type:

> "Most modern electronic equipment—for example, equipment in hospitals and cars—is shielded from RF energy. However, energy from cellular telephones may affect **malfunctioning or improperly shielded** electronic equipment."

After receiving the revisions, Wheeler replied on January 19, 1994: "I feel much better about what you have written. Thank you for your responsiveness to my previous concerns. What you have drafted is a responsible statement."

TRUTH AND LABELING—II

Carlo was high and inside the CTIA's top echelon in those early days. When he made some recommendations to Wheeler, on November 29, 1994, the CTIA president quickly turned them around and sent them on to the FDA, stating that "We have reviewed the SAG recommendations and agree that these measures would serve the public interest by providing uniform information about wireless instruments." On December 9, Wheeler passed along those recommendations—word for word—to Dr. Elizabeth Jacobson, deputy director of the FDA's Center for Devices and Radiological Health. (He did not include in that letter the title Carlo chose for his interim report: "Potential Public Health Risks from Wireless Technology.")

The recommendations of Carlo's SAG are worth noting:

1. Adopt standardized labeling of wireless instruments.
2. Develop standardized information for dissemination to member companies and to the public.
3. Adopt an industry-wide instrument certification program that requires certified phones to meet all appropriate standards.

The main reason these statements are worth remembering is that Wheeler promised the FDA they would be implemented by the first quarter of 1995. But it never happened. In the summer of the year 2000 the industry made the same promises, one more time. And that proved to be a major PR coup—a press-agent's dream—as the industry's promises in 2000 received favorable and largely unquestioning media coverage. Once again, the industry promised that the public would soon be given access to the electromagnetic emission levels of the mobile phones they were buying and using.

What this meant, in Washington's ever-practical PR and political terms, was that the industry's savvy leadership had bought the industry valuable time: a six-year grace period. Surprisingly, the FDA seemed not to have noticed, as its officials repeated the promises in the year 2000 as if they were a significant breakthrough. And even more surprising was the fact that most of the mass news media seemed unaware that the industry promises that were being portrayed as newsworthy in the year 2000 had actually been made back in 1993—but never fulfilled.

PART TWO

CHAPTER FOUR

FOLLOW-THE-SCIENCE: *HENRY'S COMET*

SCIENTISTS WERE POLITICKING in the lobby and meeting rooms of the Copenhagen Sheraton Hotel in June 1994 as the members of the Bioelectromagnetics Society, a prestigious international group of academics, were meeting-and-greeting and promoting their latest efforts. George Carlo, who was new to this field of science but already a major power in it, because of the $25 million he would be doling out to researchers, was there to learn and become known to others. Meanwhile, many of the famous scientists in attendance wanted to get to know him, too. This was the dynamic in play when Carlo was approached by a fellow whose name-tag identified him as Dr. Henry Lai of the University of Washington. Dr. Lai said he had "some very interesting data" that he wanted to show the new head of the cellular phone industry's $25-million Science Advisory Group (SAG)—an operation that, by virtue of its many dollar signs, was already known by all of the conference attendees. Dr. Lai and his colleague, Dr. N. P. Singh, had conducted a series of experiments on rats that had been exposed to radiation similar to the type of radiation that comes from the antenna of a cellular phone. Lai said the work showed that cell phone radiation causes damage to DNA in human blood cells. That certainly got Carlo's attention—a finding that cell phones *can* cause genetic damage would be a major development.

But Lai also got Carlo's attention with a pointed remark: He said he had sent Carlo data in a proposal in December 1993 and never received a response—and that, Lai said, made him question whether Carlo was really serious about following up on meaningful research of findings that cell phone radiation could pose a genuine health risk. He seemed to be hinting that Carlo might be an industry lackey who didn't want to do real independent research. Finally, Carlo remembered: In December 1993 he had indeed received an unsolicited research proposal from two scientists at the University of Washington. They had wanted to do a follow-up study of genetic damage as measured through a test system called the Single Cell Gel (SCG) assay. But since the SAG program was just beginning, its scientists had not finished laying out their research agenda and they were nowhere near ready to evaluate proposals. So when the proposal arrived, Carlo did not read it but had just tossed it into a file for future review by his Toxicology Working Group, which would be evaluating research for possible funding. Now the fact that he hadn't acted on it promptly was being interpreted by Henry Lai as an indication that Carlo wasn't conducting a serious scientific effort.

The initial work of Lai and Singh had been funded through a general grant from the NIEHS and was an add-on to work they were doing on the biological effects of 60-hertz power-line exposures. There was no question that their motivation now was to use their initial findings on cell phone radiation, which suggested genetic damage, to attract funding for more research.

Lai had arranged to conduct a presentation of his findings in private meeting room with a slide projector. He had invited Carlo and representatives from Motorola: Dr. Quirino Balzano, the company's top scientist; and Chuck Eger, the attorney in Motorola's Washington office who Carlo worked with. Frankly, the presence of the Motorola people bothered Carlo:

In Henry Lai's view, I was apparently on the same side as Motorola—just an industry guy—so he saw no conflict in having us at the meeting together. Politically, he probably reasoned that I had control of the research money and that Motorola had control of me.

At this point, I do not believe it mattered to Henry where the funding came from—either from me and the SAG or from Dr. Balzano and Motorola. He had no way of knowing that the tension in the room had little to do with the science. Dr. Balzano and I had gotten off to a very rocky beginning from the day my appointment was announced by Tom Wheeler. In 1993, at a scientific conference in Los Angeles, Balzano and I had a very contentious breakfast meeting where he explained to me in no uncertain terms that he, Balzano, was going to control the direction of research into this issue, and that I should go along to get along. He made it clear that he had all the muscle of Motorola behind him and that I would not survive in this job without his support. I told him that I would do my job the way I thought it should be done, and that he should do his. As long as the science was sound, we would have no problem.

But our relationship never warmed—in fact, it became quite tense. In Copenhagen, just the night before Lai brought us all together for his slide presentation, it even got personal—over, of all things, a cappuccino. A group of us had gone to dinner and Balzano was among those at the table. As an outsider in this field, it was important for me to develop relationships with the major players and get a feel for the science and the politics of this field. I believe Balzano wanted me to remain an outsider. After dinner, I ordered a cappuccino. Dr. Balzano tried to embarrass me in front of the group by saying that in his home country, Italy, cappuccino was a drink for women—at breakfast. He said that men did not drink cappuccino—and certainly never *after dinner.*

I never forgot that dig. I was setting out in a new arena in which I would be surprised, time and again, by just how small big people can sometimes be.

• • •

Henry Lai led us into a small viewing room in which he had set up a projector with his slides in the carousel. There were a few chairs lined up in front of the screen, on which he had the introductory image already projected. Eger and Balzano sat in the chairs. I stood against the back wall. The room was darkened. Nobody said a word during Henry's presentation. All you could hear, aside from Henry's commentary, was the whirring of the fan on the projector.

Henry wasted no time jumping right to the results of his experiment, saying simply that he had conducted a series of studies that showed DNA damage in rat brain cells exposed to microwaves in a long, rounded cage known as a wave-guide. He said he had evidence of both single- and double-stranded DNA breaks, and that he needed funding to continue the work. In the first five minutes, he had given us both barrels from his scientific shotgun.

I did not know enough about the intricacies of the experimental approach he used to ask probing questions—indeed I had never before heard the term "comet assay." I learned that week that it was a new test for genetic damage, at the time still in a developmental stage. When I asked Henry if he thought this was suggestive of a danger to cell phone users, he said it was too early to tell, but that he believed in his data. At the time, I didn't know what to make of Henry or his assay, but I knew that DNA breaks would indeed represent a serious health threat.

The Motorola guys sat stunned at first. From the beginning of the cancer scare, in 1993, the best thing that the industry had going for it was that there was no real, solid scientific data showing a problem. Most of the scare was based on speculation and inference, in the industry's view. Now, the Motorola guys realized they might be facing something very different: a reputable and respected scientist raising questions based on real laboratory data. They gathered themselves and began to grill Henry about his interpretations. Henry's experiment exposed the whole body of rats to microwaves in the wave-guide. That's hardly relevant to cell phone radiation to a human head, they said. That, Henry replied, is why he needed to do more research funded by our SAG.

I asked Henry if he would welcome me to his laboratory to learn more about what he had done, and he said he would. I left the meeting and looked for a phone.

Carlo's toxicology program colleague, Ian Munro, would be able to evaluate the significance of the findings. It was noon in Mississauga, Ontario when Carlo reached Munro at his office near Toronto.

Ian was initially skeptical, and cautioned that this could be an attempt to create a sense of urgency about Lai's proposal so that we would fund it. Ian knew about the comet assay—the technical name for it was the Alkaline Single Cell Gel Microelectrophoresis assay, or SCG assay. But his information was that it was an in vitro *test (done on cells in tubes and dishes), and not an* in vivo *test (done on live animals). Ian's best information was that this use of the comet assay had yet to be validated, meaning that the scientific community had not fully accepted it as a tool for testing. Lai and Singh had taken a test developed for use in petri dishes and test tubes and adapted it to live rats. In the best of circumstances, adapting an assay to another type of application is difficult. Ian believed that because a test that was designed to be* in vitro *was being done* in vivo*, there was no reason for us to rush to act on it. His view was that we needed to look closely at what they had done, and then to refer the matter to our Toxicology Working Group for advice on how to proceed. From the research perspective this made sense; however, from a public-health perspective it was a little more complicated.*

Our job in the SAG was to evaluate all new data and judge whether there should be a public-health intervention in which the government or industry would take some public step, ranging from a consumer notification to a warning label to a product recall. We had already conducted a review of the science that was available prior to 1994 and saw no reason for concern. However, these new data from Dr. Lai's lab—although they had not yet been peer-reviewed—presented us with the first of many tests of the public-health protection model we were following. Lai's early data raised questions about what we should do with unpublished data that had not yet been peer-reviewed but which raised concerns that could have an impact on consumers and public health. These same questions arose throughout the program, every time a new study showed up with what seemed to be a finding of some health risk. The protocol we followed was influenced by the work of Dr. Lai and his aggressive approach to us.

• • •

As soon as he returned to Washington, Carlo conferred with Ian Munro and Bill Guy on how they should deal with findings such as Lai's that were alarming but had not been peer-reviewed. Because the data were in a research proposal—a request for funding—they needed to treat it as a possible indication of a public-health problem. The fact that the data were unpublished in a scientific journal was not as important as the fact that the data had not been peer-reviewed—an indispensable part of the scientific process. Carlo, Munro, and Guy decided that they needed to have some type of peer review of the data before proceeding, but since it had not been their study they couldn't send it to their Harvard-based Peer Review Board. So they decided that the next-best step would be for Guy and Carlo to visit the Lai laboratory, see firsthand what had been done and how it had been done—and then refer the research proposal to their Toxicology Working Group, which was under Munro's direction. Lai was traveling in China during the month of August, and so the visit of Carlo and Guy to the lab at the University of Washington was scheduled for the first week in September 1994.

Meanwhile, Carlo, Guy, and Munro decided that all questions about the necessity of any form of public-health intervention would be deferred until they had a better chance to review the procedures and do an independent analysis of Lai's experiment. One thing was clear: If there turned out to be proof that DNA was broken, or that there was genetic damage, it would indeed be a serious, significant finding.

INTERLUDE: A PRIMER ON DNA BREAKAGE AND GENETIC DAMAGE

DNA can become damaged and cease to function as it should. There are three general categories of DNA damage, defined by how the damage occurs.

1. **DNA damage can occur when body cells are replicating.** Living mammals continually need new cells because old ones wear out, or because the body's organs are simply growing. As the cells replicate, each chromosome makes an exact copy of itself. The cell then splits into two cells, with each carrying one copy of the chromosome. (When sperm and egg cells replicate, each ends up with two copies of chromosomes—creating twice as many opportunities for DNA to become damaged.) DNA in cells is most susceptible to damage during this replication process.

2. **DNA damage can also occur when cells are not replicating, but are merely functioning.** Sometimes the chemical bonds within the cell become weakened and they break; it happens as many as 10,000 times per day in each cell. Sometimes one molecule simply bumps into another as they move about in the cell, and pushes the molecule off of the DNA, causing a break in the double-helix staircase; it happens about 100 times per day in each cell. Sunlight causes DNA damage in another way—by causing the double-helix staircase to adhere to itself. Base substitution damage occurs when an outside influence alters the sequence in which DNA chemicals bind with each other—changing the way the cell functions. Finally, when a cell dies, all of these errors occur as the cell spontaneously degrades.

3. **New DNA damage can occur when cells are seeking to repair old DNA damage.** The ability of a

body to repair DNA damage is the key to all life—because in every person's life, damage occurs and is repaired millions of times each day. However, if the repair mechanisms are impaired, the repair process itself can cause new damage. And the increase of DNA damage can cause health problems.

Here is a practical illustration of how DNA damage can change the way a cell works. Let's assume that we have a simple sentence that represents a DNA chain—for instance:

THE FAT CAT SAT ON A HAT

Now, let's assume we have a correct DNA sequence of six genes (making six amino acids)—each with three bases. In this sequence, our sentence becomes six three letter words that properly reads:

THE FAT CAT SAT ONA HAT

But, if there is a base substitution error, for example substituting a B for the H, then the DNA sequence will, in effect, send this message:

THE FAT CAT SAT ONA BAT

That message would cause the cell to do something different than it normally would, because the HAT gene has been changed to a BAT gene.

Now, if there were a frame shift mutation, with the S in the fourth word deleted, the characters of the genetic message would shift and become:

THE FAT CAT ATO NAB AT_

In, this message the first three genes are sending the correct message, but the last three are sending a message that is unintelligible.

Finally, if there were a double-strand break in this DNA segment, we would have something like this:

THE FAT CAT ATE ON A BAT

Cell repair mechanisms might try to stitch these together—but that would only compound the problem, by forming a DNA chain that might read:

THE FAT CAT ATE ABAT

or

THE FAT CAT ATE ABATON

With a nonsense verse such as in this example the result just seems silly; but with living cells, each error becomes magnified to the point where it can cause serious health problems. With any of these damaged DNA sections, different amino acids would be formed than those intended by the original sequence, thus altering the body's chemistry.

CHASING THE TAIL OF THE COMET

During the month of August 1994 Carlo prepared for his upcoming visit to Lai and Singh's University of Washington laboratory by reading the available scientific literature about the comet assay—a relatively new experimental technique that was developed in the 1980s that had not yet been validated.

The Single Cell Gel assay is commonly called the "comet" assay because the scientists doing the experiment wind up looking through their microscope at something that resembles a scene astronomers might see in a telescope: fragments of DNA that look like a comet with a tail trailing behind. The key, then, is to accurately measure the length of the "tail" of the "comet." A scanning device called the Komet Imaging Scanner that would accurately measure the length of the comet's tail had been invented and was manufactured in Liverpool, England. It was used in conjunction with a sophisticated computer program. The accurate measuring of the tail is the key to the experiment: The longer the tail, the greater the damage that was caused to the DNA by some outside influence—in this case, microwave radiation.

The theory behind these comet assay experiments is that dangerous exposures would cause DNA to break or would alter the cells' ability to repair DNA that was already broken by other exposures—maybe sunlight, alcohol, caffeine, or nicotine. When DNA base pairs are broken, the segments become charged either positively or negatively. Like charges repel and opposite charges attract, so when an electric current is run near a microscope slide with brain cells that have been exposed, the current causes the charged broken fragments of the DNA to migrate or move. The negatively charged segments move through pores in the preparation of cells toward the positively charged anode of the electric current. Under the microscope, the trail of broken DNA that migrates due to the electric current looks like the tail of a comet. In a given time period, the longer the tail of the comet, the more DNA damage. A brain cell that has no DNA damage would have no tail.

The practical implementation of the test was quite a bit more complicated. It was developed to be used with living blood cells growing in culture. This was because the steps that had to be taken to prepare cells for the analysis were both time-consuming and time-sensitive. Moving through the necessary steps in a controlled time frame was difficult. Lai and Singh's attempted adaptation to living mammals was a scientific stretch but it was motivated by the presence of Dr. Singh, who was an expert in the assay.

• • •

In September 1994 Bill Guy met me at Henry Lai's office at the University of Washington in Seattle, where Bill was also Emeritus

Professor. Henry introduced me to Narendrah Singh, known as N. P., his co-investigator. N. P. was a slight man who looked every bit the laboratory scientist—and I liked him immediately. We walked together from Henry's office through a series of long white hallways and down a back stairway into the building's basement. Henry explained to me that space at the university was at a premium and that he needed more research funding to be able to compete for more space. Historically, this laboratory, under Dr. Guy and his students, Dr. C. K. Chou and Dr. Lai, had been one of the most eminent in the world of bioelectromagnetics. I was pleased that Dr. Chou, of the City of Hope National Medical Center in Los Angeles, who was also doing contracted work for our SAG, had come to Seattle to join us that day.

To be honest, however, I was shocked at how sparsely equipped the laboratory was where these experiments had been done. It appeared that Dr. Singh's space was a storage area that had been converted to an office. I could see why they needed funding. Drs. Lai, Singh, and Guy showed me the wave-guide system they used to expose the rats they had studied. These wave-guides were mesh tubes, about 2 feet long and maybe 8 inches in diameter. The system had actually been developed by Dr. Guy for a large study that he had done with Dr. Chou involving whole-body exposures of rats and the incidence of tumors, funded by the U.S. Air Force. (I took mental note that the rat study by Chou and Guy was cited by Dr. Mays Swicord, then of the FDA, as one showing that microwaves increased the risk of tumors in rats. Dr. Guy, however, did not share that interpretation. In a meeting with Carlo, Guy, Munro and the Interagency Government Working Group, Dr. Swicord said he had obtained his interpretation of increased tumors by adding together the numbers of malignant and benign tumors in the study. This was contrary to the analysis protocol that had been used in the study.)

Lai and Singh showed me how they conducted the experiments. I noted that the rats were in a wave-guide where their entire bodies, not just the heads, were exposed to radio waves. I also noted that Dr. Singh had made all measurements of the comet tails manually, using a laboratory measuring stick similar to a ruler. He showed me the procedure he followed: Looking through the microscope, he first made a visual judgment about where the comet body ended and the tail began, and also where the tail ended; then he placed the measuring stick beneath the lens, visually estimated the length of the comet tail, and wrote down the length as he had measured it visually. He did not use the Komet Imaging Scanner that I had read about in the scientific literature. The reason, Lai and Singh told me, was that they had designed these as pilot studies, just to see what would happen. A high level of quality control was not a requisite and expensive new equipment was not purchased, mainly because they believed they would later receive funding necessary to do the experiments more rigorously.

As I watched Dr. Singh demonstrate his method of testing, I was troubled by several procedures I observed:

First, there was a concern about the scientific blinding during the crucial measurement of the comet tail. As he did the experiment, Dr. Singh said he was indeed scientifically "blinded," the lab term which means that when he put the slides under his microscope, he did not know which slides were from exposed rats and which were not. But he also told me that, as a knowledgeable scientist, he could tell just by looking at them through the microscope which slides had been exposed to radiation and which had not. Thus, the protocols of proper scientific procedure would note that despite all of Dr. Singh's best intentions and efforts, the possibility existed that an observer bias could be introduced that would compromise the findings.

Second, there was a concern about the fact that the limited laboratory space might have affected the rats' physiology before they were killed. In what scientists call Good Laboratory Practices, lab animals are caged in areas far removed from the place where the animals are killed and the tissues processed. Live animals can smell the blood and tissue of the dead animals. This causes anxiety that can lead to physiological changes in the animals. And that, in turn, would have to be accounted for in the experiment analysis. While Dr. Singh was giving us a tour of the lab, Dr. Chou noted that the rats had been slaughtered in the same room in which the others were caged, and he said that it would be important in the next round of experiments to keep them separated.

The third concern was with the time that elapsed during the preparation of the rats' brain tissue. When the animal is killed, there is spontaneous degeneration of DNA—a process that begins immediately after death. It is critical, therefore, to make sure that each animal is processed in the same amount of time—from the end of the microwave exposure through the harvesting of the brain cells and preparation of the slides. The greater the amount of time during this process, the greater the DNA damage that would appear in the tissue of the brain cells that had been exposed to radiation compared to the tissue of the unexposed rats. And that, in turn, would likely be interpreted as greater damage caused by the exposure—when in fact it would probably be due to the greater length of time. It was clear to me during my visit that no logs had been kept to monitor the exact time that elapsed after each rat's killing and before the harvesting and fixing of the brain tissue. So it was unclear whether time was a factor in the results.

I noted that Drs. Lai and Singh used methods that differed from those of other scientists who had pioneered this procedure. They ran the current for twice as long (60 minutes) as other scientists whose work had been published. The stain they used on the slides to make the DNA appear flourescent was different from what I had seen in the literature. Not being an expert in this procedure, I couldn't evaluate the impact of these changes.

After the presentation by Drs. Lai and Singh, we convened in a conference room and I asked, point-blank: "Do you interpret these findings as indicating that cellular phones are dangerous?"

They both replied that they were concerned about the findings, but that until follow-up research was done, they did not know what this meant for people using the phones. Their bottom line was that more research needed to be done, and on that we agreed. The complication was that our Toxicology Working Group did not recommend that we do the as-yet unvalidated assay that Lai and Singh had used. I told Drs. Lai and Singh that I agreed this needed to be followed up, and that I could be getting back to them soon. Looking back, it is clear that they interpreted my comments as saying we would be funding them soon.

• • •

Following my visit to the University of Washington, I conferred with Bill and Ian, this time with a new sense of urgency. I was not sure what Lai and Singh's data meant with respect to cellular phone users, but I did know that we now had in our possession data that some might interpret as serious enough to require some type of public disclosure.

We accelerated the review by the Toxicology Working Group, and also referred the data to our genetic Toxicology Working Group. The toxicology experts came back to us with a straightforward assessment: because of concerns about the exposure methods and other test procedures, we believed the data were not interpretable with respect to cellular phone use. More work was needed. We took that information and asked Peer Review Board members for guidance. They came back to us with these recommendations: (1) We shouldn't repeat the Lai and Singh work until the appropriate exposure systems, which we had under development, became available; and (2) we should await the scientific validation of the SCG assay technique so that we could be confident that the methods and procedures used in this test would be recognized as authoritative.

During this time I continued to work with Drs. Lai and Singh on this project. But it was three years before we completed our review and funded their work. By then our relationship was severely strained as they blamed me—not the peer-review process—for the delay.

• • •

Carlo wrote briefly about Henry Lai's single-cell gel experiment and findings in the SAG newsletter. CNN's Steve Young noticed the piece and called Carlo for an interview. In his discussions with

Young prior to the interview, Carlo sensed that the FDA and Motorola had "spun" the reporter—Young's take on what was happening was that Carlo wasn't moving fast enough, that the FDA and Motorola thought Carlo should have been moving quickly to fund studies. Motorola did not like what the SAG was doing overall, and saw this as an opportunity to levy some criticism while appearing to be constructive. In Young's piece, both Motorola and Dr. Swicord, then at the FDA, were critical of the deliberate approach Carlo and his SAG advisers were taking; both called for studies to be commenced immediately.

By contrast, in Carlo's televised sound bite, he said, "In science, the most important thing is to be right, not hasty. Because there are fundamental problems in the data, it is important to validate this assay before moving forward. This is common scientific practice."

It was not a matter of happenstance that the CNN piece turned out as it did. Motorola executives had worked hard to make it happen, beginning with two carefully crafted media strategy memos in which the company's stated goal was an effort to have "war-gamed" the Lai-Singh findings. The War-Game memo and a Question and Response memo were drafted in December 1994 by Norman Sandler, one of Motorola's top corporate communications executives. They were sent to Michael Kehs, who was working on the Motorola account for the Burson-Marsteller public relations firm in Washington. Sandler began by referring to two of his bosses who apparently had ordered the effort to create just the right corporate spin—Albert R. (Rusty) Brashear, Motorola's corporate vice president and director of corporate communications, and Bob Weisshappel, executive vice president and manager of Motorola's cellular subscriber group. In a journalistic public service, a copy of the memo, was obtained and published by the trade publication *Microwave News,* in January 1997.

MEMORANDUM

To: Michael Kehs	Date: December 13, 1994
From: Norm Sandler	Re: Revision of Lai-Singh Materials

Rusty just had an animated telephone conversation with Bob Weisshappel, who was . . . adamant that we have a forceful one- or two-sentence portion of our standby statement that puts a damper on speculation arising from this research, as best we can.

I tried to do that in the latest proposed revision of the standby statement, but offer this new, somewhat strengthened version of the second paragraph for consideration:

"While this work raises some interesting questions about possible biological effects, it is our understanding that there are too many uncertainties—related to the methodology employed, the findings that have been reported and the science that underlies them—to draw any conclusions about its significance at this time. Without additional work in this field, there is absolutely no basis to determine whether the researchers found what they report finding—or that the results have anything at all to do with DNA damage or health risks, especially at the frequencies and power levels of wireless communication devices."

. . . we should be able to say that . . . [the Lai-Singh studies]: Were not conducted at cellular frequencies, so are of questionable relevance; Run counter even to other studies performed at 2450 MHz, raising possible questions about the findings.

. . . I think we have sufficiently war-gamed the Lai-Singh issue, assuming SAG and CTIA have done their homework. We may want to run this by George Carlo and fill him in on the contacts we've made.

Excerpts of

Confidential Working Draft #3 – 12/13/94

Developments in Radiofrequency/DNA Research: Position Paper

Question and Response

How can Motorola downplay the significance of the Lai study when one of your own expert consultants is on record telling Microwave News that the results—if replicated—could throw previous notions of RF safety into question?

It is not a question of downplaying the significance of the Lai study. In his comments to Microwave News, Dr. [Asher] Sheppard [a Motorola science consultant who was also on the Harvard-based peer review board] raised the key question: Can this experiment be replicated and interpreted? We will have to wait and see. Until the results of follow-up studies are in, any conclusion about the significance of this study are pure speculation.

There is another reason to caution against jumping to drastic conclusions—the hypothesis doesn't square with human experience. If cellular radio signals could cause DNA damage, we would expect to see increased cancer rates among people exposed to RF energy. But there is no evidence to suggest this is the case.

What studies can you cite to prove RF energy doesn't affect DNA?

We have identified at least 18 published studies of animal and cell cultures exposed to electromagnetic fields (microwave frequencies, RF, and ELF) that show no effect on DNA.

Action Planned

In addition to the response materials already prepared by the SAG (see attached copies), we will work with the SAG to identify appropriate experts to comment in general on the science of DNA research, in addition to any experts SAG may be able to recommend to publicly comment on one or both of these particular studies.

Media Strategy

It is not in the interest of Motorola to be out in front on this issue because the implications of this research—if any are industrywide. Therefore, we suggest that the SAG be the primary media contact followed by the CTIA. It is critically important that third-party genetic experts, including respected authorities with no specific background in RF, be identified to speak on the following issues:

* Problems with the . . . studies.
* The health implications of DNA single-strand breaks.

We do not believe that Motorola should put anyone on camera. We must limit our corporate visibility and defer complex scientific issues to credible, qualified scientific experts. We have developed a list of independent experts in this field and are in the process of recruiting individuals willing and able to reassure the public on these matters. SAG will be prepared to release Munro–Carlo memos, which touch on key points made in this material.

• • •

I had never seen or even heard about the existence of Motorola's war-gaming memos until I read about them in the Microwave News. *Even though they mentioned me, Ian Munro, and our Science Advisory Group in the memos, they had never told me what they were up to. But at one point, Michael Kehs, who was handling Motorola's PR at Burson-Marsteller, did ask me for a list of scientists they could have the media contact to talk about this. I told him that no one who was involved in the SAG program should be talking to the media on behalf of Motorola. I never gave him a list of names. I was convinced that we were doing the right thing scientifically by waiting until we had appropriate exposure systems and a better understanding of what the findings would mean biologically from tests done on the work of*

Lai and Singh. The scientific uncertainties made it difficult to make a case for public-health intervention based on these data, and it seemed to me that we had a scientific consensus among the SAG, the Toxicology Working Group, the Peer Review Board, and the special group Dr. Munro had convened to look at the comet assay for a scientifically careful and deliberate approach.

Looking back, it is clear to me that the Lai and Singh experiments sent ripples of fear through the industry. The deliberate approach we were taking was scientifically sound, but politically not expedient for the industry—and not the approach Mays Swicord wanted the FDA to take, although it is not clear to me whether that position was his alone, or that of the FDA at the time. The questions raised were not going to go away quickly, which is what the industry would have hoped for. To his credit, Tom Wheeler and the CTIA—at least publicly—backed our decision to be deliberate.

CHAPTER FIVE

FOLLOW-THE-POLITICS: *THE CARE AND FEEDING OF WATCHDOGS*

THE RINGING OF THE FAX machine startled Carlo as he sat at his desk shortly before 9:00 A.M. on September 1, 1994. The message that fed slowly out of the machine startled him even more.

It was a handwritten memo from Tom Wheeler, addressed to his inside team—CTIA vice president Liz Maxfield, vice president for communications Ron Nessen, and George Carlo. It was written at a time when the General Accounting Office (GAO), an investigative arm of the Congress, had been questioning whether Carlo's scientific study was really independent, since Carlo's effort was being financed by the industry and he, in fact, had been hand-picked by the industry to run the research. The Democrats were the majority party in Congress, and Rep. Edward Markey (D-Mass.), the chairman of the powerful House Subcommittee on Telecommunications and Finance, had scheduled hearings to look into the matter.

In this memo, Wheeler had sketched out some thoughts that amounted to a lobbyist's version of a battle plan. Wheeler wanted to mobilize, for political reasons, the Harvard connection that Carlo had carefully arranged for scientific reasons.

Carlo glanced at the very short memo that was written in a shorthand form, and was startled by the first notation. It said they should set up "a pre-emptive strike on Markey." Wheeler went on to write that he wanted Harvard's Dr. John Graham to accompany Wheeler and Carlo to see Markey as soon as possible, in order to convey the message that Harvard is assuring the independence of the industry-funded research project.

Wheeler had other ideas for ways the industry could use Harvard's ivy for its own corporate camouflage. "Immediately following" the meeting with Markey, Wheeler wrote, "we should take Graham to GAO." The message for that maneuver in this lobbying flanking mission would be, as Wheeler jotted it without punctuation, "How dare you suggest this isn't fully independent." Finally, if the CTIA really felt it needed what Wheeler called "a fallback," he raised a question: why not just "send all cash through Harvard?"

I was a bit stunned to see the suggestion of having a 'pre-emptive strike' on a member of Congress. I really couldn't imagine throwing around the Harvard connection—and trying to use a distinguished scientist such as Dr. Graham for political leverage. It seemed crude and frankly, unacceptable to me. So was the idea of using Harvard University as a way of passing the industry's money to the scientists to make it all somehow look independent. I had no doubt that our research would be independent. Well, as sometimes happens in government and industry, this was one of those instances where nobody said anything about the memo we all received. And thankfully, nothing was ever implemented.

• • •

But another very different meeting did occur—between the top brass of the cell phone industry and the top brass of what seemed to be most of Washington's alphabet of agencies: The EPA, FCC, FDA, NCI, NTIA, NIEHS, NIOSH, and OSHA (Occupational Safety and Health Administration).

This was a major event for Wheeler's industry. It was vital that they convince the government that the industry was doing the right, scientifically responsible thing—that independent research was being conducted and that the health and safety of the public were being protected. Otherwise, the government regulators might decide to step in and aggressively oversee (the lobbyists would say: overregulate) this industry. After all, since the government had never insisted on advance testing to certify the safety of these radiation-emitting phones aggressive action now could lead to any number of scenarios, all restrictive from the industry's perspective.

The meeting in May 1993 was in an auditorium at the huge FDA headquarters, in the Washington suburb of Rockville, Maryland. Carlo's role in Wheeler's well-orchestrated presentation was to outline the entire scientific effort. Both Wheeler and Carlo left the auditorium convinced that the government now believed that industry could be trusted to protect the people who were buying their product. The government would not aggressively intervene. Mission accomplished.

• • •

The public had every right to expect that the federal government was exercising its proper oversight role in regard to the safety of cellular phones. But from the time that wireless phones were first introduced into the marketplace to the time in the year 2000 when 100 million Americans were using the devices, both the legislative and executive branches failed to do everything that they could—and should—have done to safeguard the citizens they are obligated to serve.

At a time when diligent oversight was needed, Congress and the federal agencies were too often shortsighted. And at times, the bureaucrats clearly found it more convenient to just look the other way.

On November 21, 1997, Rep. Markey, by then the ranking minority member of the Commerce Committee's Subcommittee on Telecommunications, Trade, and Consumer Protection, sent the FDA a list of questions concerning the agency's oversight and investigation of the health effects of wireless communication. It took a while for the agency to reply, but in a letter dated January 14, 1998, the FDA indeed responded: "As you will note from the answers provided below, there is no new information indicating that use of cellular phones is a human health risk. It is our hope that ongoing studies and those planned by a number of other organizations will shed some light on this important health issue." (That answer led many to anticipate that the FDA would say something different—and mainly, take action that was different—a year later, after new findings showed a significant risk of health effects. However, those who anticipated such a responsible response from the FDA would eventually be disappointed.)

People everywhere who hold cell phones against their skulls several times a day would have been very interested in hearing what the regulators had to say in answer to the congressman's fifth question: "What provisions have been made to ensure that there are long term monitoring studies of users of wireless telephones? Are specific studies in place? Who is conducting such research?" Carlo's Wireless Technology Research organization had, after all, said that long-term testing is essential.

But the FDA officials chose to reply by defining their own responsibility narrowly, bureaucratically: "As you know, we have no regulatory authority to require manufacturers of electronic products to conduct specific long-term studies."

That FDA response contrasts significantly with the statement by which the FDA's Dr. Elizabeth Jacobson referred to her agency in a letter to Carlo on March 13, 1997: "As the lead federal agency charged with regulation of radiation-emitting consumer products, the Food and Drug Administration has followed the progress of your research into possible health effects of wireless technology with great interest."

And despite the occasional flurries of news media coverage that raised questions about the safety of cell phones, there was no public outcry—and certainly no special interest—that could have turned these concerns into a cause. As a result, this public-health issue never developed into a political imperative that could have galvanized senators and representatives into action. And that meant there were no assertive congressional investigations or demands that, in turn, would have jolted the government agencies out of their bureaucratic inclinations to do little and do it slowly.

But there was one exercise that the federal agencies all seemed willing and eager to perform—and that was turf building.

So the FCC took the position that devices that emitted low-power radio frequencies should be excluded from the provisions of federal emission guidelines that required a device to emit no more than 1.6 watts per kilogram (W/kg). (Indeed, lawyers for the Swedish cell phone manufacturer, Ericsson Corporation, for example, had written the FCC in September 1994 urging the adoption of a categorical exclusion for low power handheld devices—and added that the FCC should not adopt any rule that requires that tests be conducted on mobile phones to determine if they comply with government emission standards.)

The FDA objected to that. Then, on October 11, 1994, the FCC took a most interesting, if not persuasive, position in a letter to the FDA's Dr. Jacobson. Richard M. Smith, the FCC's chief of the Office of Engineering and Technology assessed "the requirement in the exclusion clause that a 2.5 centimeters separation be maintained between a radiating device and the 'body' of the user"—and came up with this joint government and industry interpretation: "The interpretation of the working group and of Ericsson is that this separation was not meant to apply to the head of the user of a hand-held device."

In other words, the user of a cell phone should keep the phone 2.5 centimeters away from his or her *body* because there might well be a health risk if it is held any closer. Yet the FCC had no concerns about having that same phone and its antenna held directly against the *head* of the person using the phone.

Just six weeks earlier, an FCC official had been quoted by the Bloomberg news service as saying that the FCC is not in the business of doing basic biological research to ascertain how cell phones might affect the brain. "We don't have the authority to do that sort of thing," Dr. Robert Cleveland, an FCC environmental scientist, was quoted as saying. "The FDA is more in line to do that kind of thing."

But the FDA's Dr. Jacobson persevered, replying to Smith that "several recently published scientific studies indicate that cellular telephones . . . can be used in a manner that may induce local SARs [specific absorption rates] that exceed 1.6 W/kg in the heads of users." Therefore, she wrote, the FDA believed these devices should be "'certified' by their manufacturers not to exceed the local SAR limits . . ."

And so, in 1994, the FCC yielded to the FDA and cell phones were required to be certified by their manufacturer that they would meet these standards. But five years later, ABC News' *20/20* news magazine program would report that a number of phones it had tested were still emitting radiation in excess of that 1.6 W/kg standard. So in the year 2000, the cell phone industry announced with great fanfare that it would begin making the SAR levels available to the public, to assure that its phones were all finally in compliance with the government guideline. Over at the FDA, new officials in charge greeted the industry's belated action with commendation, as if they had truly lost their institutional memory. And as things usually happen in the public-events food chain, the news media, in turn, greeted the industry's announcement as a major news-making event—showing that the members of the media, too, had lost their occupational memory.

A PROTEST TO THE MEDIA

On August 29, 1994, CNN aired another report on the safety of cell phones by correspondent Steve Young. The following morning the report was repeated on CNN Headline News. Later that day, August 30, Wheeler faxed to Carlo a copy of a letter of protest he would be sending to CNN President Tom Johnson. Wheeler's "Dear Tom" letter complained that the CNN coverage was "not complete and factual." He complained that coverage of studies by Dr. Om Gandhi that showed higher radiation absorption rates than were shown in earlier studies had been "sensationalized." Wheeler also took issue with the correspondent's statement in the piece that just one research grant had been processed. And the industry's top lobbyist actually launched a staunch defense of Carlo's program. "If we had rushed into research without the total scientific impartiality and peer review represented by the Harvard-peer review process, that research would have been suspect as being too industry-controlled,"

Wheeler wrote. "Instead, we did the right thing with independent experts following all the appropriate scientific procedures. To take a gratuitous shot as a throw-away line in an already misleading report compounded the misimpression given to your viewers."

One day after Wheeler faxed his letter to Johnson, the CTIA's Ron Nessen received a report of the results of a focus group, conducted by a public opinion research firm, which seemed to confirm that Wheeler's instinct and concerns were on target. The report, from public opinion analyst Neil Newhouse, reported on the findings of two focus groups, composed of a total of 21 persons outside Philadelphia who had been shown the CNN story and then asked about what they had seen.

"After viewing the news segment, fully 62% of the participants had the impression that the level of radiation emitted by cellular phones is above federal safety standards, while just 29% believed them to be below federal safety standards," wrote Newhouse. The report listed a number of comments made by those who watched the CNN segment, ending with one person who said: "It does not indicate the manufacturers are taking any precautions . . ."

Nessen forwarded copies of the focus group report to Wheeler, Liz Maxfield (CTIA vice president in charge of overseeing the health and safety issue), and Carlo. His handwritten note attached said the focus group study "proved exactly what we feared."

CHAPTER SIX

FOLLOW-THE-SCIENCE: *HEART-STOPPING PHONE CALLS*

ONE BY ONE, IN THE EARLY 1990S, the reports made news around the world. And one by one, they were filed away in the offices of the FDA. Heart pacemakers were occasionally going out of whack—after the cardiac patients used their cell phones. Heart defibrillators were malfunctioning—after cell phones or two-way radios were used in close proximity. There was even a report about a motorized wheelchair suddenly starting up and hurtling its occupant over a hillside in Colorado, causing a broken hip—apparently triggered by radio waves.

None of these were about cancer—but all of them were about electromagnetic interference. The possibility that cell phones were interfering with medical devices such as pacemakers, devices in wide use, quickly became a prime focus of FDA concern. It also quickly became a focus of Dr. George Carlo's research—much to the consternation of some top officials of the cell phone industry which was funding his program. Carlo's work on the pacemaker–cell phone problem produced several lasting results: Working in direct conjunction with the FDA, Carlo organized the research and produced the recommendations that pioneered a lasting solution to a serious public-health problem. But by not taking his cues from the CTIA and by operating with what he saw as scientific independence, Carlo infuriated Tom Wheeler, creating a lasting distrust that would never be repaired. That rupture in their relationship would cause Carlo considerable grief in the future.

When Carlo went to the 1994 Bioelectromagnetics Society annual scientific meeting in Copenhagen, Denmark, he heard presentations of three studies that suggested cellular phones had the capacity to interfere with implanted cardiac pacemakers. Many types of heartbeat irregularities can be corrected with pacemakers, devices that transmit controlled electronic pulses to the sinus node of the heart to regulate beats. Modern pacemakers are usually implanted in the collarbone region of the shoulder, and have two wires that go to the heart, carrying the electric signal.

In Italy, researchers tested European digital cellular phones in a laboratory and in patients with pacemakers. Thirty pacemakers were placed near cell phones in the laboratory and tested for any tendency to produce irregular pacing and inhibition of pacing. When the phones were in close proximity to the pacemakers—less than 10 centimeters (close to 4 inches)—interference of some type was found in half of the instances. In 101 patients with 43 different pacemaker models from 11 different manufacturers, the antennas of two different phones were placed on the chests of the patients near the pacemaker. Interference of some type was seen in 26 patients when the pacemaker was programmed to its most sensitive setting.

In Switzerland, researchers reported on their studies of digital handheld cellular phones in 39 pacemaker patients. They tested whether exposure to the cellular phones, placed directly over the pacemakers, caused the pacemakers to either speed up or stop. They saw both types of interference in a few patients, and called for more detailed studies to see how widespread this type of problem was in patients with pacemakers. (The tests, since they used live cardiac patients, required a series of ethics reviews before they could be done.)

In Australia, Dr. Ken Joyner—who later was hired by Motorola—had studied ten patients with pacemakers, and reported that adverse outcomes occurred when the phone antennas were held 20 centimeters (about 8 inches) directly above the pacemaker.

While the studies were not definitive, they raised important questions about the safety of pacemaker patients using wireless phones. The FDA queried the CTIA about the significance of these findings. The CTIA had told the FDA that the technology being used in Europe and Australia was significantly different from the cellular phones in use in the United States. And, interestingly, industry officials further told government officials that the phones that showed interference in the international studies were using digital signals, and that only analog technology was in use in the United States. (Analog phones have a continuous wave, similar to an FM radio; digital phone signals are pulsed.)

Carlo and his colleagues later learned in discussions with FDA scientists that FDA officials had concluded that if the problem with pacemakers was confirmed, they could place a moratorium on the introduction of digital cell phones in the United States until a solution to the problem was found. This would have been a devastating blow to the wireless phone industry at this stage. The industry was anticipating the launch of new digital technology in the United States and the White House and FCC were anticipating earning billions of dollars for the federal treasury in the auctions for the Personal Communication System frequency bands. The reason the FDA did not see interference with pacemakers as an immediate threat was that the industry had led the FDA officials to believe that digital phones were years away from being in use in the United States. Carlo was surprised to discover that the FDA officials were unaware at that time that digital cellular phone systems were already being run on a pilot basis in the United States. Moreover, one of the pilot programs was running right there in Washington, D.C.

This underscores again the degree to which the FDA was limited in its understanding of the technology, politics, and economic interests of the wireless industry it was responsible for regulating.

• • •

With the first public reports that mobile phones could be interfering with hospital and medical equipment, two phone manufacturers—Ameritech and Bell Atlantic—responded quickly and independently, without waiting for their trade association to act. They established programs to work on the problem with medical officials.

Bell Atlantic Mobile (BAM) hired the powerhouse Washington public relations firm Hill and Knowlton to develop and implement a program focusing on the early complaints about medical devices. Hill and Knowlton outlined its proposal in a July 1994 memo to BAM: "The overall objective is to educate hospitals, the healthcare community and consumers on the electromagnetic interference issue, especially as it affects cellular phone usage. At the same time, the program will position BAM as a responsible corporate citizen, source of accurate information and provider of a safe and important product."

The Hill and Knowlton memo to BAM was a mix of all-too-rare corporate responsibility and all-too-common PR-speak. It recommended "anticipating and addressing appropriately and as positively as possible all aspects of the issue proactively," and came up with a specific bold proposal: The company would issue a "statement establishing and describing a BAM 'Swat Team' that can go to hospitals on-site and assess and analyze situation . . . and can identify any potential problem areas and offer solutions." The company strongly supported continued research into the problem. But the PR firm's suggested key message points put the onus on the hospital equipment manufacturers to shield their equipment from cell phone interference: "The focus of this issue should be on safe and adequate shielding of existing and new medical devices," the PR firm's memo said. ". . . The manufacture of many types of medical devices that may be affected by EMI [electromagnetic interference] is essentially unregulated for adequate shielding. . . . BAM is opposed to medical facilities banning of cellular and other wireless devices as a solution to EMI issues . . . "

In a memo reacting to the proposed campaign, BAM executive Brian Wood wrote: "Lots of good stuff. Thanks." He then noted that the company felt its industry trade association wanted to move too slowly and BAM needed to do the right thing, right away: ". . . The CTIA plan seems too long term for us. Hospitals need help right now. They can't trash their current equipment. It seems . . . [to] make more sense to provide technical support for them today."

But at the CTIA, Bell Atlantic Mobile's effort to do the right thing was not viewed as the right thing to do. CTIA vice president for communications and public affairs Ron Nessen sent a copy of the Hill and Knowlton memo to the three insiders: Tom Wheeler, Liz Maxfield (CTIA vice president overseeing the cell phone health and safety issue), and George Carlo. And Nessen attached to it a handwritten note dated July 22, 1994, that said simply:

"Bell Atlantic follows Ameritech in charting an independent course on EMI—more defensive, more apologetic than our position. Not a good trend. Ron."

• • •

Three months later, in September 1994, Dr. Roger Carrillo, a heart surgeon at Mt. Sinai Hospital in Miami, faxed to Carlo's office preliminary results of a study he was conducting on 59 patients who had pacemakers. He had heard of Carlo's program and was looking for funding to continue his work. Carrillo had a test system that allowed him to assess interference in pacemakers from digital cellular phones of the type that were being pilot-tested in the United States. He was using telephones he had gotten from Motorola, whose scientists had programmed the phones to the "test" mode—able to send signals but not receive them. He programmed the pacemakers to a high sensitivity setting and conducted tests on the 59 patients with a number of different phones. When the phone was held directly over the pacemaker, he saw interference in 21 patients. Of the 170 tests he conducted, 39 showed interference; in 21 of those tests the most severe form of interference occurred—the pacemakers stopped functioning. The interference affected 19 different pacemaker models. Dr. Carrillo saw no interference when the phone was held to the ear in the usual talking position. Patients in the tests volunteered to participate, and doctors were at their side throughout the procedure. No patients experienced harm during the testing.

Contrary to what the FDA had been led to believe, the problem was no longer limited to the type of cellular phones used in Europe and Australia. Dr. Carrillo's findings were with phones already being used on a pilot basis in the United States—and that would soon be in wide use throughout the country. After receiving the data from Dr. Carrillo, Carlo asked Dr. David Hayes of the Mayo

Clinic in Rochester, Minnesota, to review the work. Hayes confirmed that the data raised important safety questions. He had seen similar interference in his patients as well. Carlo immediately notified the FDA that there was now evidence of a potential problem with pacemaker interference in the United States. One week after he'd gotten the first data from Dr. Carrillo, Carlo and his advisers were briefing the government on the results and their implications. They met at the FDA's offices outside Washington, D.C., in suburban Rockville, Maryland. Chuck Eger, the attorney for Motorola, attended the meeting as the representative of the cellular phone industry.

• • •

Jeff Nesbit (our Science Advisory Group, liaison with the FDA), Chuck Eger and I arrived at the FDA headquarters in mid-afternoon. From the moment that we walked into the crowded conference room it was clear that the FDA was taking this meeting very seriously. There were a number of government officials sitting at the large table wearing their Public Health Service uniforms—much like the one that I used to see Surgeon General C. Everett Koop wear on TV—white, with gold stripes indicating rank and medals for service commendations. The room was silent as we walked in. Unlike other meetings, where small talk had to be quelled to get the meeting underway, this group of about 15 was very definitely ready for urgent business. Dr. Elizabeth Jacobson invited Nesbit and me to sit at the head of the conference table. Eger took a chair away from the conference table and to my right.

As the FDA's senior person in attendance, Dr. Jacobson opened the meeting by introducing me and saying that I had some important information about pacemakers and cellular telephones that could lead to government action. I presented the four tables of data we had received from Dr. Carrillo, I remember thinking as I spoke—with everyone in bright white uniforms, silently weighing every word and piece of data—that this seemed like a military inquiry.

There was very little discussion about the data during or after my presentation. The FDA had data from their laboratory that had shown the same type of interference. This was no longer just a theoretical problem—it was affecting Americans now, and the FDA had been caught in a complacent lull, believing that digital phones were a long way off. I could sense the group was looking to me and to Dr. Jacobson for direction.

Dr. Jacobson spoke of how the clinical data we had presented would take precedence over the laboratory data that the FDA had generated. She said that it would be much better if there were more data on patients to judge how big the problem was and to identify solutions. She reviewed with us the regulatory authority that the FDA had exercised over the pacemaker manufacturers and stated that the pacemaker industry would have to be involved. She also made it clear that the FDA had not exercised any regulatory authority over the cell phone industry—and that whatever that industry decided to do voluntarily would determine what regulatory steps the FDA would have to take.

The message was impossible to miss: If the industry did the right thing voluntarily and supported the scientific work that the FDA needed to identify a solution to the problem, then the FDA would not take any formal regulatory steps.

We agreed that the clinical study approach was in everyone's best interest. Dr. Jacobson asked me to oversee the process of doing the research. I agreed.

When we left the meeting, Nesbit and I had a short meeting with Eger to review what had been decided and Eger agreed to notify the industry, both Motorola and the CTIA. On his cell phone, I overhead him saying, "During the meeting, Dr. Elizabeth Jacobson expressed the critical need for clinical data on the phones being used in the United States. The FDA asked the SAG to become involved in the process, and a plan to conduct a large clinical study is now in place."

• • •

The issue of pacemakers and cell phones quickly caught the attention of politicians in Washington. On October 5, 1994, the U.S. House of Representatives Government Operations Committee's

Subcommittee on Information, Justice, Transportation, and Agriculture held a public hearing. There was general agreement that research was needed, and the study being planned by the SAG was an adequate beginning.

Carlo and his advisers developed a protocol for the clinical study. It was to be a cooperative effort of the FDA, the SAG, the cellular phone industry, and the pacemaker industry. Scientists from the Mayo Clinic, Mt. Sinai Hospital, the New England Medical Center in Boston, the University of Oklahoma, and The George Washington University participated. A separate group of clinical experts was asked to specify which types of interference would be considered most dangerous, and for the study, interference was classified according to their recommendations.

The FDA and the scientists working under the SAG jointly analyzed and interpreted the data. This was a highly unusual degree of cooperation, made possible because the FDA viewed the scientific process as independent enough from the industry to present no conflict of interest problems. Indeed, that very independence seemed to displease Tom Wheeler. He expressed his displeasure to Carlo about the fact that the government could see the data in the early stages and his people could not.

• • •

I was very proud of the way Wireless Technology Research (or WTR; the legal entity established by CTIA and the SAG to oversee the research and surveillance efforts to assess the health impact of wireless technology, created in response to General Accounting Office recommendations) and the FDA worked so closely together in overseeing the pacemaker interference study. It was the type of working relationship that would not have been possible if the FDA did not believe that the WTR was independent of the industry, and it was positive and constructive.

When we got to the point of having the first data available for analysis, we set a meeting to have Dr. Hayes come to Washington and present the findings to us at FDA headquarters in Rockville. Word of the meeting somehow reached Wheeler, and I got a phone call.

"George? Tom. I hear you have a meeting scheduled next week to

go over the pacemaker interference findings. Somehow, we were not notified about the time and place."

This was an unusual call from Tom. He almost never got involved in the hands-on work. Something was going on here—he was giving me the opportunity to blame the CTIA*'s not being invited on an administrative snafu. I had no problem with him knowing about the meeting—everything we did in the* WTR *was open—but to have* CTIA *involved would have been a mistake.*

I responded, "Yes. Dr. Hayes is going to bring his data, and we are going to work together on the analysis. But it would not be appropriate for either CTIA *or* HIMA *(the pacemaker manufacturer trade group) to be there. That is why the meeting is just with the investigators."*

"We funded the study, George. We should have a representative there. Once Hayes talks about the data, it will be all over town. We need to know." Wheeler was becoming testy.

"I can't agree to that, Tom. I'll give Jo-Anne [Basile] a briefing after the meeting. That's the best I can do."

"I'll talk to Liz Jacobson about it, then," was his response.

"Fine, Tom, but I won't agree to it, even if she does."

The phone call ended. Our meeting at the FDA *took place without a* CTIA *representative present.*

THE MAYO CLINIC'S PACEMAKER/CELL PHONE STUDY

The purpose of this study was to assess the prevalence of interference and the potential for serious clinical risk to patients with permanently implanted pacemakers from exposure to cellular phones.

This clinical study was conducted simultaneously at the Mayo Clinic, the New England Medical Center, and the University of Oklahoma Health Sciences Center. In the study, 980 patients with pacemakers were tested for interference from five types of wireless telephones—one analog and four digital in use in the United States. The telephones were programmed to the test mode, to transmit at maximum power to simulate a worst-case scenario. One digital telephone was also tested during actual transmission. Patients were monitored with an electrocardiogram while the telephones were put through a series of maneuvers directly above the pacemaker. Interference was measured and judged as to whether it would pose a danger to the patient. A total of 5,553 interference tests were done.

The findings were straightforward: Interference of some type was seen in 20 percent of the tests. The amount of interference was different for each type of telephone. Less than 3 percent of analog phone tests showed interference, while nearly 25 percent of the digital phone tests did show interference. Importantly, the incidence of interference was higher during an actual call as opposed to when the phone was in the test mode. The incidence of interference was significantly more frequent when the telephone was placed over the pacemakers compared to when the telephone was in the normal talking position against the ear. And most important: Pacemakers that were equipped with additional filtering on the wires that enter and exit the device had the lowest incidence of interference.

The significance was clear: The study unequivocally confirmed that digital wireless telephones had the capacity to interfere with implanted cardiac pacemakers when they were placed directly over the pacemaker. Conversely, the incidence of interference was minimal when the telephone was placed at the ear in the normal talking position. Models of pacemakers with filters on the leads seemed immune to the interference.

The findings led officials from the cell phone and pacemaker industries and the government to implement short-term and long-term solutions to the problem of interference between digital wireless telephones and implanted pacemakers.

In the short term, it was decided that manufacturers of pacemakers and cell phones would inform their customers that pacemaker wearers should keep their wireless phones more than 6 inches away from their pacemakers. They should not keep the phone in the "On" position in a breast pocket, for example. They should keep their cell phones as far away from the pacemaker as possible.

Long-term, it was decided that all pacemakers would be manufactured with lead-wire filters that shield against the interference.

Looking back, the solution of the pacemaker–cell phone problem was a rare example of how well the system can indeed work. Just 24 months had elapsed between the time the problem was first identified and the time it was solved.

• • •

During the early years of the WTR program, we looked at solving the pacemaker interference problem as not only an important accomplishment for public health, but also the model for how the WTR program was supposed to work.

Namely: Through our surveillance effort, we had identified a potential problem of interference that could seriously affect patients who use cell phones. We went to the responsible government authorities and notified them of the problem. They asked us for help in the science and we obliged. The research we oversaw was focused and served to identify the scope of the interference problem and offered both long- and short-term solutions. We had the government's endorsement of our recommendations. The cell phone industry and the pacemaker industry were both very much involved and cooperative by the end. In effect, the problem was solved.

• • •

But the industry saw it differently. They saw Carlo's SAG (which became the WTR during this period) as being extremely independent, which, in the view of the industry, meant "beyond control." The interactions between Carlo and the CTIA became very strained through the process. Wheeler's vice president in charge of overseeing Carlo's work had several heated discussions with him about not only the cost of the work but also the recommendations that the WTR was about to make. While Carlo took input from CTIA as he always had, the industry was not happy with the recommendations and believed that the problem should be solved entirely by the pacemaker industry putting filters on pacemakers.

When the analysis of the pacemaker interference data was completed, it was clear that there was a problem with interference. All types of phones were involved, but digital phones caused more interference than analog phones. The working group that included the Food and

Drug Administration, WTR, and the investigators drafted a series of recommendations.

I called Jo-Anne Basile to invite her to the office to get a briefing on the draft recommendations. She came with Art Prest, CTIA's representative on the protocol committee.

Sitting at the conference table outside my office, we had a very heated exchange.

I had pointed out that our recommendations put the short-term burden for intervention on the phone industry to notify customers who have pacemakers to keep the phones away from the pacemaker. It put the long-term burden on the pacemaker industry to design pacemakers that are immune to interference.

Basile seemed quite agitated, as she told me, "You are out of step with where the industry is on this. You have not consulted us. You are always off on your own."

"Jo-Anne, these recommendations came from the group and they include the FDA. This is what needs to be done." I said.

"This is not the phone's problem. This is the pacemaker's problem. We are not responsible." Now it was clear that she was giving me the industry's position.

"Let's be realistic. They are not going to recall a million pacemakers. The responsibility has got to be shared."

"It is not up to you to tell us what our responsibilities are," she said. It was clear to me that they felt I was not acting in the industry's best interest. I felt I was acting in the best interests of the science and the public.

• • •

The successful collaboration of the FDA and WTR on the pacemaker interference issue ended in a unique afternoon of dueling press conferences. In September 1996 the WTR convened a scientific colloquium to publicize the results of the research and the joint FDA-WTR recommendations.

Following the colloquium, the WTR held a press conference to give the media the opportunity to ask questions of the investigators and government officials who participated. Drs. Hayes, Wang, and Carrillo were present, representing the investigators. Don Witters and Paul Ruggera represented the FDA. Dr. Kok-Swang Tan represented the Canadian Health Protection Branch. The press conference was held in a small room adjoining the Capitol Hilton's Federal Room, where the colloquium was held.

During the WTR press conference, which I moderated, after a number of questions and answers I noticed a gathering in the hallway outside. I walked to the door and was shocked to see Jo-Anne Basile holding her own impromptu press conference. The CTIA had opted not to participate in ours.

At Basile's press conference, the CTIA issued a press statement from Tom Wheeler in which he spoke of the cellular phone industry's rapid and responsible approach in identifying the problem and proposing a solution. Wheeler's press release did not mention the WTR—nor, of course, did it mention Carlo, or the vital role each played in finding the lifesaving solution.

CHAPTER SEVEN

FOLLOW-THE-MONEY: *CAUGHT IN A LITIGATION VISE*

IN APRIL OF 1996, Carlo and the Wireless Technology Research (WTR) program seemed to be under attack from all sides. The CTIA had stopped funding the WTR's research—which forced Carlo to notify scientists around the world to stop working on the projects for which they were already under WTR contract. That caused Carlo's scientific colleagues, who had agreed to do work for the WTR but had not been paid in months, to criticize openly Carlo's program and his leadership of it. The trade press and mainstream media ran articles laced with criticism of Carlo and the WTR. The media reported that three years after the effort had been launched with much public fanfare, there was still no significant research finding to parade before the public.

And just when it seemed that things could get no worse for Carlo, they did. He found himself suddenly caught in a potentially calamitous vise of litigation. The pressure would tighten intensely, threatening to wipe out all of his assets unless he abandoned his independent research effort. Carlo soon learned that one who was applying the pressure was Tom Wheeler.

The WTR, and Carlo personally, were named as defendants in two lawsuits that had been brought against the mobile phone industry, its top executives, and its chief researcher. The CTIA was named as a defendant in both suits, as were Tom Wheeler and Ron Nessen; the WTR and Carlo were also named as conspirators with the industry in both cases. For Wheeler and Nessen, the legal cases were an irritant but hardly a crisis; their personal liability would of course be safeguarded by the indemnification of the CTIA. But Carlo discovered, much to his surprise, that the cell phone industry was taking the position that it was not intending to stand behind him financially—even though he was being sued only because he had been doing research that was funded by the industry.

The first lawsuit was brought by Debra Wright, an Arizona mobile phone company employee who had developed a brain tumor. She filed the lawsuit in Illinois, alleging that the industry was involved in a global conspiracy to keep information about the dangers of cellular phones from consumers. She said her bosses had encouraged her to use a cellular phone for a great deal of time. She had alleged that while the industry knew of the dangers in the 1980s, she was never warned; and she alleged that the actions of Carlo and the Science Advisory Group (SAG) were promoting the continuation of the conspiracy to defraud the public of information that could protect them.

The second lawsuit was brought, also in Illinois, by Gerald Busse. He was not ill, but he alleged that the epidemiology studies being conducted by the researchers under contract to the WTR constituted an invasion of privacy and were in violation of the FDA's rules forbidding unlawful human testing. The reasoning behind the claim was that phone records of cellular customers were being obtained and reviewed without their express permission in order to compare heavy users with brain tumor and other records. Thus, the suit maintained, the privacy of all phone customers was being violated. As in the Wright case, Busse also charged that the WTR and Carlo had conspired with the cell phone industry to withhold health-risk information from the public.

• • •

The total claims in the two lawsuits amounted to more than $100 million in damages. And since Carlo did not have adequate insurance coverage of his own to cover any legal obligations, he was facing a catastrophic crisis. All of his personal assets—including his house and car—were exposed and at risk.

I suddenly realized that I was in danger of losing everything. I needed some type of indemnification from the industry to protect me from the potential liability of the lawsuits if I was going to stay on with the WTR *program and move forward with its research. I was far too vulnerable for my comfort.*

When the first lawsuit was filed naming me as one of the defendants, I talked about the need for indemnification with Liz Maxfield, the CTIA*'s vice president who was overseeing the health and safety issue. She was a very decent woman and she assured me that the* CTIA *would indemnify us—the* WTR *and me, personally—against any harm or financial obligation arising from the lawsuits. I took her at her word.*

Late in 1995, Tom Wheeler and I engaged in a heated exchange about our litigation costs and CTIA*'s refusal to bear the burden of them. As Liz Maxfield and I left his office, she walked me down the long hallway and into her office, and closed the door. I could see that she was clearly upset—and worried.*

"Please do not let him know I promised you the indemnification," she said. "I need this job. I have a new baby daughter. I will do everything I can to help you, but please protect me." And I did protect her throughout the proceedings; I am discussing the incident now only because she left the CTIA *a couple of years later.*

• • •

The defense of the Wright and Busse lawsuits had cost the WTR over $800,000 in outside legal fees through the end of 1996. In addition, hundreds of hours of WTR staff time were spent not on necessary scientific work but on preparing documents and doing other work to support WTR's legal defense, including preparation for depositions. These litigation-related costs had been paid out of the WTR's operating budget because there was simply no other source of financing for them. And as the WTR's funding and staff time were diverted from other research efforts, the trade press stepped up its hammering of Carlo about the slow pace of the research.

On April 7, 1996, Carlo received a phone call from Wheeler. He wanted to meet with Carlo the next morning—and in a precedent-setting choice of venue, he wanted to meet at Carlo's office, in a stately old townhouse on N Street NW.

Wheeler had never been to my office before. He usually stayed in his own domain. This was highly unusual.

Tom arrived at 9:00 A.M. *sharp. Collette Herrod, our receptionist, let me know that he had arrived, and I went down the stairs to the first floor reception area to meet him. He was already halfway up the stairs, and was amusing himself looking at the pictures of our office outings that adorned the wall. His picture was there—from 1994, when he had a dark moustache; he was on a boat we had chartered for a joint SAG-*CTIA *fishing trip, wearing a baseball cap and smiling. Those were happier days for us all.*

I led him to my office and closed the door behind us. He looked at my diplomas and certifications that hung over the couch opposite my desk. He had a bemused look when he caught the autographed picture of Casey Stengel hanging over the fireplace, the Babe Ruth memorabilia, and the pictures of Joe DiMaggio and Mickey Mantle. He sat on the couch, facing four pictures of me with George Washington University basketball players and mentioned that his daughter, Nicole, played basketball. I sat in a wooden arm chair next to the couch.

I asked, "What can I do for you, Tom?"

His response was short and to the point. "I can no longer support the idea of the WTR *as an independent group and give you indemni-*

fication in these lawsuits. Our lawyers have advised me of this, and there is not much I can do about it."

I listened as he continued. "You know, George, when it was the old way, the SAG with you working on the inside, it was better. That is what we should be going back to."

"So what do you suggest, Tom?" I asked.

"We can indemnify you only if you are an employee of CTIA. We can have you run this whole program out of the Cellular Foundation, our nonprofit arm, and as an employee we can rid you of the pressure from these lawsuits." He went on to say he'd give me a title of vice president.

He had me caught in a vise; he knew it, and now he was tightening the grip. He knew I did not have insurance coverage for the type of conspiracy claim that had been brought by the plaintiffs. While the WTR was covered, personally I was not. He knew I was in such a desperate bind that the temptation to just say yes and relieve these legal and financial burdens on me was compelling—to say the least.

But I told him that the independence was a critically important aspect of the WTR program. I went on to say that, besides, the FDA would never go for the idea of me running the research program from inside the CTIA. He told me that he could handle the FDA.

I wish I could say that this was all strictly noble and altruistic on my part. But the truth is that I also had a card to play—and I figured it was an ace in the hole. I knew that the WTR Audit Committee, with some members appointed by the CTIA and some by the WTR, had secretly secured a deal with Wheeler in which it received complete indemnification by CTIA, yet still remained independent of CTIA.

Truth is, I found out about the Audit Committee's indemnification deal only by accident. One day I simply asked the Audit Committee's Chairman, Jerry Polansky, what he thought I should do about the indemnification problem. I also decided to ask if he was worried about his own liability. He told me straight out, "I have no problem. We're fully indemnified."

"How? By whom?" I asked.

He told me that Steve Hooper, the president of AT&T Wireless, one of Wheeler's handpicked members of the Audit Committee, had told Wheeler they needed indemnification as legal protection. Of course Wheeler had no choice but to oblige this request from the AT&T Wireless president, a member of the CTIA board.

I said to Wheeler: "Tom, if I can have the same indemnification agreement as the Audit Committee, then I will take my chances with the independent WTR. They are the same as we are, and it seems that you should be able to offer us the same without your lawyers having a problem."

He was stunned, but he was also stuck—and he knew that, too.

Our meeting ended abruptly when I told Tom that working for CTIA would not be an option for me, and he walked out.

• • •

For six months, Wheeler and Carlo hardly spoke to each other. They dealt with each other mainly through intermediaries as the complicated indemnification principles and details were being negotiated. Then, in October 1996 Wheeler telephoned Carlo.

His demeanor was upbeat; that had now become a warning sign to me. He said, "Let me buy you lunch. I'll come by at noon, and we can walk to Herb's."

My legal counsel in this negotiation, Linda Solheim, had given me a heads-up that Wheeler might be in a mood to call me. She had been negotiating these issues with Wheeler's attorneys, and they had finally put together a proposal. The CTIA was concerned that the WTR's legal strategy was more time-consuming—more expensive. Indeed it was, because we felt we had to put the WTR's program on trial to show it was indeed independent of the CTIA.

I told Tom I would meet him at Herb's. Although the restaurant was located in the Holiday Inn, just a half-block from my office, I had never eaten there. When I arrived, Wheeler was waiting for me out front. We went in together. It was clear that Wheeler was a regular; the host knew him, and took us to a table in the back.

We sat down and as I was picking up the menu, Wheeler said: "They have the best Cobb salad in town."

"Fine. Two Cobb salads."

We got down to business. By now Wheeler had heard from his lawyers the WTR's arguments regarding the need for indemnification, but I went over them again, just to make sure. The WTR could not do anything or agree to anything that even appeared to be inconsistent with our independence. I expressed my view that giving CTIA control of the litigation process would compromise our independence. In both the Wright and Busse litigations, the WTR and CTIA had taken different approaches—we had different goals and sensitivities—and the WTR needed to continue to have its own counsel. Lawyers representing CTIA would not be able to represent WTR without some conflict. The WTR should not have to use research funds to address cash-flow needs associated with litigation expenses; those costs had to be covered by CTIA. But a promise to pick up the costs was not good enough. Given the problems CTIA had in collecting funds from industry, I was concerned that once our research had ended, it would be even more difficult obtaining the money to pay for litigation expenses that would likely be ongoing. We needed a signed, written agreement.

Wheeler laid out CTIA's concerns, which I had heard before. The CTIA did not want to agree to anything that was outside the realm of common indemnification clauses. The commonly accepted procedure would give them control over the litigation process if they were paying—including choice of counsel, legal strategy, and settlement parameters. They viewed blanket indemnification as a blank check to plaintiffs that would bring more lawsuits upon the WTR as a way of getting to the industry's money. He also reiterated what he had said to me in a very contentious letter that past May: the legal fees that WTR had accrued were exorbitant and the legal strategy we had followed was questionable.

The Cobb salads hadn't even arrived and already we were at another impasse, and at each other's throats. Then Wheeler surprised me. He handed me a memo from his attorneys, Philip Verveer and Jennifer Donaldson. It was stamped "confidential," although I knew that as soon as he gave it to me, the attorney-client privilege was waived, and it was no longer really confidential.

"We have a solution. Why don't you look this over and see if we can agree on it." He dug into his salad.

The memo, referring to legal-fees indemnification, was very carefully worded. It said that they, Wheeler's attorneys, believed the scientific independence of the WTR could be maintained while permitting CTIA the ability, subject to a standard of reasonableness, to participate in the WTR's defense of litigation. It laid out a series of recommendations that in one way or another did the same thing—it gave CTIA control over litigation costs, and therefore indirect control over the WTR's approach to litigation defense.

As we discussed the memo, I agreed to send a letter directly to Wheeler's attorneys with my response. We had finished our salads and were actually still talking civilly. That, at least, was rather nice. Wheeler picked up the tab.

We left and walked back down N Street toward my office. As Wheeler left me in front of my office and continued down N Street toward his, I shook my head. Although I could not agree with the recommendations in the memo, we were finally making progress. With my attorney, I drafted a response and a counterproposal.

The Compromise

Carlo's letter to Wheeler's attorneys became the framework for the agreement on how the WTR's litigation costs would be handled. After negotiation over a few minor points, an agreement was struck.

CTIA would not have control of the WTR's litigation.

WTR would not have blanket indemnification coverage, but insurance coverage.

CTIA agreed to replenish the WTR escrow fund with $800,000 to cover the out-of-pocket litigation costs that WTR had already paid. CTIA made a claim for reimbursement to its insurance carriers.

The bottom line: The negotiated Memorandum of Understanding between the WTR and the CTIA granted Carlo the same indemnification coverage, through a series of insurance policies, as the members of the Audit Committee had. The pressure of that legal and financial vise on Carlo had been eased, just in time.

• • •

In April of 1996, Carlo did not know what the outcome of these troubling lawsuits would ultimately be. As it turned out, in both suits, Wheeler, Nessen, and Carlo were dismissed as defendants because they were not residents of the state of Illinois where the court had no personal jurisdiction over them. The WTR remained in the Wright case until it was dismissed in 1997. The basis for the dismissal was another Illinois case, *Verb v. Motorola,* which had determined that, as regards health risks of cellular phones, state law was superceded by federal law. The court reasoned that the FDA was guarding consumers against the health risks of cellular phones, and that any actions by the state court would be redundant as long as the FDA was doing its job.

In the Busse litigation, the WTR prevailed in its arguments at the state court level that no laws had been broken and that no privacy rights of cellular phone users had been violated by the epidemiological studies. At this writing, the case is currently on appeal in Illinois.

CHAPTER EIGHT

FOLLOW-THE-SCIENCE: *INTERESTING BUT INCONCLUSIVE FINDINGS*

DURING THE EARLY AND MID-1990S, a number of new scientific studies produced alarming findings which seemed to indicate that cellular phones could indeed cause major health risks—including that of cancer. Some of the studies made news around the world; others never caught the attention of the news media, but George Carlo and his top scientific advisers investigated them all. And when they followed the science, they discovered that each of the studies had flaws of one sort or another—questionable procedures which did not mean that the findings were necessarily wrong, but which did mean that the findings could not be validated as proof that cell phones posed a genuine health risk.

As they followed the science, Carlo and his advisers found that each study, while flawed, still provided important clues for their investigation.

LEAKS IN THE BLOOD BRAIN BARRIER

The phone startled me as I looked up from the fax. It was 6:30 A.M., and I was at my desk trying to catch up on my mail and paperwork. Who would be calling at this hour? It was early March of 1994, and it was snowing in Washington.

I did not recognize the heavy accent on the other end of the line, but when he identified himself as Dr. Leif Salford, calling from Stockholm, I sat up in my chair. Dr. Salford was a Swedish physician who had done research into the effects of microwaves. I knew him by reputation. "Good morning, Dr. Salford. What can I do for you this early morning?"

"I am sorry for calling so early, I thought I would be able to leave a message."

"No problem. Just trying to catch up a bit. What can I do for you?"

He said he had heard about the new research initiative we had begun with industry funding, and he thought I would be interested in knowing of his group's ongoing work. Arvid Brandberg, head of the Swedish telecommunications industry trade group, had given him my number. I thought to myself, this must be important if Brandberg had him call me.

"For several years we have been looking at the effects of microwaves on brain function," he said. "We are now convinced that we have demonstrated a potentially dangerous effect of the radiation that could come from the antennas of mobile phones. We are consistently seeing a breakdown of the blood brain barrier following exposures that are the same as those from mobile phones. I think this is important."

The surveillance function of the SAG was my responsibility. It was my job to look at all new scientific information such as Salford's to see whether it suggested a problem that would require some type of public-health intervention. Since the program began in March 1993, I regularly received calls from investigators who believed they had work showing problems; I had to determine which had merit and which were frivolous. Most of the time, I encountered researchers who were exaggerating the significance of their findings in order to leverage funding from the SAG research program. But Dr. Salford was clearly different. We had already cited his work in our Research Agenda.

His research found a breakdown of the blood brain barrier? He certainly had my attention.

For years before I became involved with wireless phone research, I had studied the impact of chemicals on people. From my experience with herbicides and pesticides, I knew the importance of the blood brain barrier. Brain tissue is extremely sensitive to trauma from physical and chemical insults. To survive and to maintain brain function, humans have evolved specific protections for the brain. One such protection is, of course, the hard skull; another is the special filter that keeps dangerous chemicals from reaching the brain. That filter is called the blood brain barrier. Blood vessels in the head are different from blood vessels elsewhere in the body in that they have unique filtering characteristics to protect the brain from being penetrated by harmful chemicals.

If it were shown that exposure to the radiation from a wireless phone antenna causes a breakdown in this very important defense mechanism, this would be a serious problem. In turn, this could constitute an indirect mechanism for the development of brain cancer—with the radio frequency radiation (RFR) not causing brain cancer directly, but providing a pathway for other cancer-causing chemicals to damage sensitive brain tissue that would otherwise be protected.

This was, of course, all speculative. But if it proved true, it would be no small problem, because people are exposed to dangerous chemicals every day, in their workplaces and from the events of everyday life. For example, smokers have dangerous chemicals from cigarette smoke continuously circulating in their blood. When they also use a cellular phone, would they be putting themselves at an increased risk of brain cancer? Would those cigarette carcinogens, which would not have been able to reach the sensitive brain tissue had the blood brain barrier been intact, now be able to act directly on brain cells? Cancer would not be the only concern. Because brain cells do not have the types of protective defenses that other tissues in the body have, the brain would be vulnerable to other toxicity as well.

By that time in 1994 there were an estimated 90 million people using mobile phones worldwide. Even if the risk of harm was small, it could still affect millions of people.

Through his heavy accent, I could sense the concern in Dr. Salford's voice. "Dr. Carlo, I believe we should meet."

• • •

I flew to Stockholm, and during the flight read everything I could find on Dr. Salford's work and the relationship between microwaves and the blood brain barrier.

The scientific literature was replete with conflicting studies regarding the effects of microwaves on the brain. The innovative approach that had been used by Dr. Salford and his group to study the blood brain barrier, using a unique type of chemical marking, was of particular interest to me. Their concept was to chemically stain brain tissue of rats, then look through a microscope to see if there was a difference in the stain patterns between rats that had been radiated with radio waves and those that had not been radiated.

Dr. Salford met me at the office of the Swedish telecommunications trade association. He wasted no time, getting right to the details of his work.

In their experiments, Dr. Salford and his colleagues exposed rats in a chamber that guided microwaves at 915 megahertz over the rats for two hours. The animals were allowed to roam in the chamber, so the exposures covered their whole body. Each animal's brain and all other organs were exposed to some degree, but it was not possible to measure precisely how much radiation went into the brain in the experiments. The investigators had to estimate the exposure in ranges. The exposures to RFR included both continuous waves, similar to analog phones, and pulsed waves, similar to the signaling that occurs with digital phones. The specific energy absorption rates (SARs) varied between .016 and 5 watts per kilogram (W/kg). These SAR levels, although estimates, were well within the range of that emitted by mobile phones. After the two-hour exposure, the rats were killed and their brain tissue was harvested and fixed on slides with fluorescent chemicals so that leakage of protein through the blood vessels could be assessed under a microscope.

Salford and his group had found significant leakage—a breakdown—in the blood brain barrier at all SAR levels. At the higher SAR levels, above 2.5 W/kg, their data appeared to follow a dose-response—the higher the level of radiation dosage, the more severe the leakage in the blood brain barrier.

Rats exposed to estimated SARs in the range of 0.4 to 2 W/kg showed a 200 percent increase in the occurrence of leakage of protein through the blood brain barrier. Rats exposed in the range of 4 to 5 W/kg had a 500 percent increase in risk. These risk increases were statistically highly significant.

When their presentation was over, I asked, "Do you think your findings suggest the need for some type of protective intervention for consumers?"

Salford answered, "I cannot say whether this opening of the blood brain barrier is an absolute health hazard, but I don't believe it is a good thing."

"On the basis of your data, is there something we should do with consumers?" I asked.

"I don't want to make any recommendations yet about what to do with consumers."

I was not surprised by the noncommittal response. Most scientists do not want to step out of the comfortable territory that is their data; they are more comfortable saying that their findings should simply beget more study. As the meeting wound down, Dr. Salford handed me a concept paper detailing the research approach they would take if we funded them.

• • •

When I returned from Sweden, I consulted my colleague Ian Munro. His reaction was cautious: "These studies are hard to do, and they are using a new methodology. Until we have done a careful review, we should not make any judgement. This could just be an artifact. We can take a look at it as we look at the other proposals we have received."

With regard to our public-health intervention responsibility, Dr. Munro and his colleagues on the Toxicology Working Group reported back that the data did not represent a clear picture of a hazard with respect to people using cellular phones, and that the group could not see the need for a consumer intervention. They felt that opening of the blood brain barrier itself would be of little consequence—unless a person was simultaneously exposed to other carcinogens. But then again, many people are exposed each day to carcinogens—smokers, for example.

Still, given the uncertainty over just how much radiation the rats had endured, we took no action other than take note of the finding as we awaited results of other studies.

Before we got around to funding any follow-up work by Dr. Salford, the cell phone industry would limit the scope of the research it would fund. But given what we now know about the biological effects of radio waves, I believe that studies of the blood brain barrier should be given a high priority—especially since Wireless Technology Research (WTR) has developed new and better exposure systems. The blood brain barrier problem has not been pursued by any of the research groups working to assess the dangers of mobile phones. Anything that compromises the blood brain barrier provides a delivery pathway for carcinogens and other poisons to sensitive brain tissue. People, through their lifestyles, diets, and occupations, are exposed daily to poisonous chemicals. The absence of the blood brain barrier as the brain's primary protective mechanism could cause the estimated 500 million mobile phone users worldwide to be at an increased risk of tumors and of sustaining other types of brain toxicity.

FIRST BIOLOGICAL EFFECTS IN RATS

In the early 1990s the scientific thinking about the development of cancer was that it is a process involving a number of biological steps. The cancer-causing process begins when a normal cell, following exposure to chemicals or radiation, undergoes one or a number of genetic changes. The changed cell clones itself and that group of cloned cells begins to grow in an abnormal manner, thus becoming a tumor. When the growth becomes random and uncontrolled, the tumor is said to be cancerous or malignant. When the malignant tumor spreads to other tissues in the body, the tumor has metastasized.

When the first questions about cellular phones and brain cancer were raised in 1993, it was widely believed that radio waves had inadequate energy to break DNA bonds and therefore could not cause genetic damage. If radio waves from cellular phones could not cause direct genetic damage, then the reasoning was that they could not be carcinogens. At the WTR, until they had scientific proof, Carlo and his advisers were reluctant to assume that radio waves did not cause genetic damage.

Even if it were assumed that radio waves could not initiate genetic damage, then the focus of research regarding the development of cancer from cellular phones would logically be on whether radio waves, while not initiating this damage, might still exacerbate any previous damage. If a cell were mutated by another carcinogen—for example, cigarette smoke—then radio waves might play a role in promoting the development of a tumor. This was the assumption made by Motorola scientists as they moved forward with their own internal research effort through the 1990s. They commissioned a number of studies aimed at assessing the promotional effect of radio waves on cancer.

• • •

Dr. Ross Adey's research project at the Veteran's Administration Medical Center in Loma Linda, California, financed by Motorola, was the first to find biological effects from cellular phone radiation.

A respected and fiercely independent scientist who had been a leader in radio wave research for three decades, Adey conducted a series of experiments. The results from one of them in 1996 surprised the scientific community and the industry. It showed that biological effects were being induced by exposing rats, headfirst, to radio waves generated by a digital cellular telephone. Prior to this finding by Dr. Adey, there was no evidence from studies of living animals that radio waves from cellular phones had any biological impact whatsoever.

In this experiment, rats were exposed to an agent that caused genetic damage while they were still in their mothers' wombs. The offspring were then later exposed to radio waves for 23 months, beginning on the 35th day after their birth. The experimental system that Dr. Adey and his team had devised was the first in the world able to approximate the type of head-concentrated exposure that humans sustain when they use the cellular phone. With very sophisticated methods for measuring both the amount of radiation sustained by the animals and the radio waves' biological impact on the animals, the experiment was a very important step forward in the overall research process for cellular phones and health effects.

Ironically, there was a second surprise in Dr. Adey's results: The biological effect he saw appeared to be protective, rather than damaging. Rats exposed to radio waves from digital phones actually had fewer tumors than rats who were not exposed. While never claiming that cellular phones would protect against cancer, Dr. Adey was sure that his work had shown a biological effect of some kind from exposure to digital phone signals.

A subsequent experiment completed by Dr. Adey and his group in 1997 failed to show any biological effect from analog phones. With the same experimental approach, the analog phone study examined 540 rats, with the brains of the males exposed to an average radiation level of 2.3 W/kg and the females 1.8 W/kg. There were no effects on brain tumor incidence that could be attributed to the analog phone exposures. In the study, however, the highest levels of radio wave exposure were not in the brain per se, but in an area further down the back of the animal—a fact not known until after the study was completed. Thus, it was possible that there had been a difference in the amount of exposure received by the heads of the rats. And that could have contributed to the difference in the findings between the digital phone study (which showed a biological effect) and the analog phone study (which did not). Further work is clearly necessary.

Since Dr. Adey's work had shown a surprising protective effect of radio waves, we saw no need to inform the public or to take any type of public-health step. However, we took careful note that, for the first time, cellular radio waves had been demonstrated to cause biological effects. I noted that we needed to inform the industry that it should amend its public position that radio waves were biologically inactive.

LYMPHOMA IN OVEREXPOSED MICE

In May 1997 Dr. Michael Repacholi and his colleagues from the Royal Adelaide Hospital in South Australia published research about a discovery that made the cellular phone world take notice: the first scientific evidence that cellular phones could cause cancer.

In the highly regarded scientific journal *Radiation Research*, it was reported that long-term exposure to the type of radiation that comes from digital cellular telephones caused an increase in the occurrence of lymphoma in mice. The study received widespread international media attention because it was the first time that cancer had been linked to cellular phones in a well-conducted scientific investigation.

The cancer study used mice that were genetically engineered so that they were predisposed to developing lymphoma, a serious disease affecting the immune system. Experiments with these types of animals—called transgenic mice—had come into use in the early 1990s because using these mice offered advantages over other types in research. In experiments with these genetically engineered animals, the time needed to glean scientific answers about cancer induction is shortened from roughly four years to one year because the genetic alteration has made them one step closer to developing tumors than normal mice. Also, the use of these transgenic mice means that fewer animals can be used in the experiments—because the background incidence of tumors—the variable that determines how many animals need to be included in a study for it to be statistically sound—can be reduced. This makes the studies considerably less expensive than whole-life studies of normal animals. For example, a whole-life study of the development of cancer in rats exposed to radio waves would cost between $8 and $12 million; an equivalent study of transgenic mice would cost less than $1 million.

In the experiment, Dr. Repacholi and his colleagues studied 101 female mice exposed for two 30-minute periods per day for up to 18 months (or for the lifetime of the mice, if they lived less than 18 months). They compared the incidence of tumors in the exposed mice with 100 female mice who were not exposed but were kept in the same environmental conditions as the exposed mice. Radiation exposures were carefully measured in the cages that held the mice. The investigators were able to estimate that the amount of radiation that reached the mice averaged 0.13 to 1.4 W/kg. This range of radiation exposure is similar to that received by cellular phone users.

The Australian group's experiment uncovered a higher incidence of lymphoma in mice exposed to radiation than in mice that were not exposed. Exposed mice were 2.4 times more likely to have the tumors than were the control mice, and the difference was statistically significant. The investigators believed the study showed that ". . . long term intermittent exposure to RF [radio frequency] fields [similar to those from cellular phones] can enhance the probability that mice carrying a lymphomagenic oncogene will develop lymphomas." At that time in 1997 this finding was totally unexpected, based on the other science that was in hand around the world.

The science at that time still seemed to indicate that there were no genetic effects from radio waves—and it was unclear how tumors could occur without some type of genetic damage occurring first. However, the investigators hypothesized that a cancer-promoting effect could be possible without a genetic effect as drastic as actual breakage of DNA. For example, they said a small amount of low-grade radiation on a daily basis could cause changes in cells that are not readily apparent, but could still lead to the growth of cancer. In other words, they believed that exposure to intermittent radio waves for a substantial part of the lifetime of the mice could have a cumulative effect that, over time, could lead to tumors.

When the Australian study was released, I referred the report to the WTR's Expert Panel on Tumor Promotion, headed by world-renowned cancer expert Dr. Andy Sivak. Dr. Sivak's expert group identified a number of issues present in Dr. Repacholi's work that made it difficult to assess whether the study pertained to users of cellular phones and more importantly, to the safety of those phones.

First, the panel of experts believed it was difficult to extrapolate the Repacholi findings directly from mice to humans because of differences in the way mice and humans absorb RFR. In the experiment the mice were exposed through their whole bodies, whereas people who use cellular phones are exposed primarily in the head and neck. The study with these mice is particularly difficult to apply to people because no humans are presently known to carry the same gene that is activated in these mice to make them prone to lymphoma. The actual relevance to humans is unknown, and could be assessed only with further corroborative information.

Second, the Repacholi studies were of an unusually long duration, which created a new confusion and uncertainty in the interpretation of the results. Studies of this type are usually six months long—but Dr. Repacholi chose to study the mice for 18 months. He explained that he wanted his study to be powerful enough statistically to identify even a small effect from the radio waves; information from the supplier of the mice indicated that by 18 months, the incidence of lymphoma would be high enough to allow for identification of an effect that would cause just a doubling of the incidence—a very desirable statistical outcome. However, the longer study period created confusion in the interpretation of his work because the mice that had died after six months showed no difference in tumor incidence between those that were exposed and those that were not. If the study had been stopped at six months, with all of the mice killed and analyzed at that time (as is the common practice with studies of this type) perhaps the results would have shown no tumor development effect from radio waves at all—which is what was observed in those mice that actually did die after six months.

Third, as leaders in the field, the WTR's tumor promotion expert panel had completed a comprehensive review of all of the available science at the time regarding radio waves and the promotion of tumors and had concluded that the weight of the scientific evidence did not support the theory that radio waves promote cancer. The findings from Repacholi's study were contrary to the scientific consensus at that time.

We decided that the study, though very well designed and conducted in a rigorous manner, did not compel us to issue a public-health warning. The uncertainties in it led us to convene a scientific workshop on the use of these types of studies for assessing the risks of radio waves; it was held in 1998 and included representatives from government and industry. That workshop concluded that much more needs to be done before it is clear how these types of studies with genetically engineered animals can help in identifying health hazards from cellular phones.

Furthermore, because within the coming year we would complete a full battery of tests underway to look for genetic damage and the risk of tumors in cellular phone users, we were comfortable deferring any further action until the new data were available.

HEADACHES

In early 1998 Dr. Kjell Hansson Mild, from the Swedish National Institute for Working Life, and his colleagues from Norway reported that as people increased their usage of analog and digital phones they experienced a correspondingly profound increase in the prevalence of headaches, fatigue, and the sensation of warmth around the ear.

The study was triggered by an unusual number of headache complaints made to Dr. Mild's office from government employees in Stockholm who had switched from analog to digital mobile phones in early 1995. When the workers switched back to the analog phones, they reported that their symptoms went away.

To study the problem, Dr. Mild assembled an international advisory group with representatives from the United States, Sweden, Finland, Switzerland, and the United Kingdom to oversee the development of the study's protocol, the conduct of the study, and the interpretation of the findings.

The study's objective was to assess whether there were more symptoms such as headaches, dizziness, feelings of discomfort, and difficulty in concentrating among digital phone users than in analog phone users. The study was conducted simultaneously in Sweden and Norway; more than 15,000 people, randomly selected from telephone company registers, participated.

There were no significant differences in symptoms between those people who used analog phones and those who used digital phones, which had been the original hypothesis that the study addressed. However, in both the Swedish and Norwegian data, a statistically significant increase in the prevalence of symptoms was noted that corresponded to an increase in both the number of calls made per day and the total minutes on either type of phone. For headaches, dizziness, and discomfort, the symptom prevalence increased dramatically—by as much as six times—as usage increased from less than two minutes per day to more than 60 minutes per day. A similar trend was present when usage went from less than two calls per day to more than four calls per day. The authors cited leakage in the blood brain barrier as a possible mechanism causing the symptoms.

We took note of the findings in this well-conducted, peer-reviewed study as further evidence that the radiation from mobile phones caused biological effects. The public-health step that we believed necessary was informing the public, and that is the recommendation we made to Dr. Mild.

CANCER FINDINGS, 1994–1996

Throughout the WTR program, our scientists reviewed all data relevant to cellular phone health effects. Three epidemiology studies of workers exposed to RFR on the job provided indirect information. It was the only information then available on people exposed to radio waves.

In 1994 a Canadian study of electric utilities workers showed a statistically significant increase in lung cancer in those workers who had jobs that put them in proximity with mobile radio communications equipment and with possible exposure to radiation from the equipment. For those workers in the highest exposure category there were 84 cases of lung cancer—which was calculated as a risk that was 3.11 times higher than for workers not exposed. For another 27 cases of lung cancer among those who worked in the highest exposed jobs for 20 years or more, the risk was seven times higher. We did not believe that lung cancer per se was relevant to cellular phone users, but any evidence of a carcinogenic effect of radio waves was important, so we took note of it.

In 1996 a study of 246 brain tumors among a group of 880,000 U.S. Air Force personnel exposed to RFR in their jobs revealed a statistically significant increase risk as well. The risk of developing brain tumors was 1.39 times higher in those exposed to RFR as compared to those not exposed.

Finally, an ongoing study of Polish military workers exposed to radio waves in their work revealed in 1996 statistically significant increases in brain and blood cancers. The risk of brain cancer was 1.9 times higher in workers who were exposed to radio waves than in workers who were not exposed. For leukemia and lymphoma, the risk was 6.3 times higher in exposed compared to unexposed workers.

Although the risks reported in these studies were statistically significant, we considered that these occupational studies were only indirectly relevant to cellular telephone users. The studies provided useful interim data, but we believed that our own studies of cellular telephone users and our studies using exposure systems that were specifically designed for our wireless phone research would provide more useful data.

CHAPTER NINE

FOLLOW-THE-MONEY: *TURNING OFF THE SPIGOT*

IN CELL PHONES, as in all other things, efforts to follow the politics and the science will inevitably merge into a new path of political science, where the first rule of the road is: Follow the money.

And when power-politics clashed with power-science—Tom Wheeler versus George Carlo—the balance of power was inevitably held by the same hand that held the purse strings. No doubt it should have been obvious to Carlo from the start that the CTIA president would hold the power and the purse. But that truth was not evident to Carlo at the outset of his research program. In fact, he really always thought he would be able to overcome Wheeler's money-power advantage. And when he found out that he couldn't, it was too late for him to do anything about it.

In the beginning, Carlo spoke with blind faith and wide-eyed optimism that the money and the science would be just fine. When the CTIA first announced that the industry would fund the $25-million research project, a reporter at the December 13, 1993, press conference asked Carlo if he was certain that all necessary studies could be done for that amount. And Carlo replied:

> *". . . the question of 'Is this enough money?' has to do with what we find as we move forward, and the industry folks have reassured me that there will be adequate funds to do what is necessary to answer the [health] question. So, in a lot of ways, $25 million may be an arbitrary number. The point is that the industry has committed what is necessary to address the question."*

Carlo was convinced of that because, after all, Wheeler had taken a similar public line in the days right after that first Larry King show about cell phones and cancer when he sought to combat his industry's public-relations crisis by announcing a research program to reassure people that cell phones are safe. On January 30, 1993, Associated Press reporter Diane Dunston wrote of her interview with Wheeler: "[Wheeler] said he expects the research to be costly but that the cellular phone industry would pay for all of it."

But in retrospect, it is clear that Carlo's expectations were different from Wheeler's intentions. Carlo believed that the industry was serious about paying for the research that was necessary, however costly it might be. Wheeler had always seemed confident that the $15 to $25 million spent on research would solve his industry's PR problem. This difference in understanding would later become crucial as the independence of Carlo's research agenda touched off a controversy over the dollars. The controversy erupted into a full-blown crisis that dramatically changed the scope and direction of the research effort.

I had every reason to believe, based on my discussions with Wheeler and his CTIA *representatives, that the $15 to $25 million number was a general commitment, not a specific budget as Wheeler would later claim. I thought it was just common sense that no one could have estimated the precise budget costs of a program like this without first having a defined research plan. And I assumed everyone knew that we would be able to establish a budget for the program only after the research agenda was* *completed—and that we couldn't possibly function independently if the industry was going to start by giving us a precise budget.*

My experience with research on other consumer-product health questions had taught me this program would be expensive. The one long-term animal study that had been done starting in the late 1980s, funded by the U.S. Air Force and conducted by Drs. Guy and Chou, looking at whether rats whose bodies were exposed to microwaves experienced a higher incidence of cancer, had taken eight years and cost $12 million. Our program could end up including several animal studies in addition to in vitro *laboratory tests and costly human epidemiology studies. But we didn't know how much work was ultimately going to be necessary. I knew it was going to be expensive, but no one could know how expensive until the program was better defined.*

Wheeler had not only assured me that the money would be there, but he had also assured the federal government, the scientific community, and the public through his representations to them. In retrospect, I know now that I should have forced discussion of those big-picture issues at the beginning.

When we began the Science Advisory Group (SAG), there were no written contracts between the CTIA and my group. We submitted invoices to Liz Maxfield of CTIA every month; these bills consolidated the time and expenses from the more than 150 people and institutions we employed in the program. Within two weeks we would have a check from her to cover the invoice.

For the first two years of the program, I had no reason to suspect that there was any problem with money from CTIA's side. I reasoned that a several-billion-dollar industry would find a way to pay its bills, especially for something as important as the health of its customers. I had met with Tom Wheeler, Liz Maxfield, Ron Nessen, and other CTIA staff two or three times a week as we moved the program forward, and I made sure that everyone at CTIA was aware of the scope of what we were doing. I knew we did not need any surprises with regard to the money. I thought it was working very smoothly. Cash flow was never a problem for us, until Wheeler began having problems with the independence of our research agenda.

FIRST CASH-FLOW CRISIS

The first sign of fiscal trouble appeared on January 19, 1995, when Carlo presented to the industry's Joint Review Committee his 1994 summary of what had been spent. And for the first time, Carlo presented a detailed budget—his cost projections—for 1995.

The 1994 accounting had been straightforward: Through the end of 1994, while Carlo and his science advisers were formulating the research plan, the program had spent $3.8 million.

The 1995 budget seemed, to Carlo, to be no problem: For *in vitro* and *in vivo* scientific experiments and epidemiology studies that were ready to begin immediately, Carlo budgeted $10,440,600 for the year. With the CTIA's various companies contributing $5 million a year that the CTIA was collecting for the program, Carlo believed that $15 million would have been available through the end of the program's third year, in 1995. His total projected expenditures through that time were $14,200,000—well under $15 million. He had not anticipated a problem.

But Carlo's projected cost figures for the 1995 research did not match the dollars that the CTIA said were available. The discrepancy between what Carlo believed should be available and what Wheeler said was available for the research was more than $6 million.

Eventually it would turn out that $6 million contributed by the association's companies had been spent for purposes other than Carlo's research program. Before it was resolved, the controversy would become a huge problem for Wheeler within his own industry. Members of his trade association were not happy with the way their money had been spent.

"Cell Phone Contracts Canceled; Health Research Program in Disarray. CTIA & WTR [Wireless Technology Research] Accused of Mismanagement," said the headline in the May-June 1996 issue of *Microwave News*. The story reported:

> "It seems that the CTIA spent a lot of money on things like PR," said John Madrid, Toshiba's representative on the committee that oversees WTR funding. "There have been outrageous administrative costs," he told *Microwave News*. "If WTR had

> received all the money it was supposed to, it would not have a funding dilemma at this moment." According to Madrid, the CTIA has collected $15–$16 million for research, but WTR has only been given about $12 million.

And in the May 6, 1996, edition of *Radio Communications Report,* Washington correspondent Jeffrey Silva reported:

> "I question the depth of the CTIA management's involvement in health and safety programs in light of the original commitment to maintain complete separation of the industry from the research," said John Madrid, whose McLean, Va., firm helps market Toshiba cellular phones.

Wheeler declined to be interviewed for this book and through his chief spokesperson, offered a straightforward explanation. The problem was Carlo's involvement in the book, said CTIA Senior Vice-President for Communications Margaret Tutwiler. She told co-author Martin Schram that, under other circumstances, Wheeler would have been willing to be interviewed by Schram, but he didn't believe it was in the CTIA's interests to participate in any interview for a book in which Dr. George Carlo was co-author. Tutwiler said that she had recommended this position to Wheeler and he had agreed with her. Thus, she said, the CTIA's president was declining the opportunity to have his CTIA viepoint included in this book.

The funding controversy between Wheeler and Carlo became a huge problem for Carlo within his own scientific community. He was the one who had to tell prominent scientists at leading universities and other institutions that there were no funds for projects—after he'd already informed them that their projects had been approved. In some cases the staffing and work had already begun. On the various campuses, these scientists had often seen to it that the initial good news from Carlo's WTR was widely circulated among the faculty and even in the local press. Now, when the flow of funding was halted, many of them blamed the messenger who brought them the bad news: Dr. George Carlo. In retrospect that is understandable, for in his letters to the scientists Carlo wrote merely that the WTR was "restructuring" its program, as if it were all his doing. He made no mention of the fact that the money spigot had been turned off. Carlo's letter read:

> 26 April 1996
>
> Dear Dr. ________:
>
> WTR's Scientific Program has undergone an internal restructuring. This restructuring is the result of an internal WTR program assessment process necessitated by funding demands and the need to meet specific future scientific program requirements. With limited funds difficult choices have to be made. As a result, a restructuring process will be undertaken by WTR over the course of the next several months.
>
> While the internal review is progressing, there should be no further scientific work done.
>
> Sincerely,
> George Carlo, Ph.D., M.S., J.D.
> Chairman

I didn't want everyone to get the impression that I'd lost my clout with Wheeler, or that the program was on shaky ground. I didn't want them to abandon their efforts, because I really believed I'd be able to get the money pipeline flowing again so we could all do the research that needed to be done. Looking back at it, there was no reason to cover up what was really happening. I probably hurt my reputation with some of my colleagues when I was actually trying to get them their money, save their projects, and frankly, avoid displaying all of our dirty laundry before the entire scientific community.

The relationship between Wheeler and Carlo would never again be cordial or collegial. And this would be Carlo's first hard lesson into the political facts of scientific life.

This had been the first time we had the information on the full scope of the program and the money that was necessary to do the job. I had simply given the CTIA and the industry's Joint Review Committee the budget that we had developed, after laying out our research agenda. We needed $10 million that year to move forward with what we were doing. And I believed the money had already been collected from the industry participants.

So it seemed very simple and easy. Was I ever wrong!

Liz Maxfield, Wheeler's vice president, called me the day after I submitted the budget and said that Wheeler was livid. He told her I had presented budget numbers that he had neither seen nor approved—and he was blaming her for having permitted me to surprise him in front of the Joint Review Committee. She said we needed to meet, and I agreed.

When she arrived at my office she looked very upset, and that concerned me. Over the two years of the project, Liz and I had become friendly and often talked about things other than business when we met—our children, our families, and her hometown of Austin, Texas. That day, she was all business.

She told me that CTIA could not cover the cash flow that I had presented to the Joint Review Committee. She said our research program had to be changed to accommodate a $5 million cash flow for the year, instead of the $10 million we budgeted. She gave me an explanation that had to do with something about the calendar-year cycle we were on for the program and how that was different from CTIA's fiscal-year cycle, which coincided with the federal government budget cycle. It didn't make sense to me. This was not like her—the demeanor, the words. I knew something more was up, but had no clue what it was. It was only later that I learned that the money that had been collected for our program had already been spent inside the CTIA.

When we ended the meeting, I told her we would have to meet again when I had a better idea about what cash-flow flexibility we would have for the various projects we already had underway. I also made it clear to her that I viewed this cash-flow problem as CTIA's, not the SAG's. Our job was to do the science; CTIA's job was to collect from the industry the money that was needed to do the science.

As the dispute between Carlo and Wheeler became known throughout the industry, the government, and among the trade-press journalists, the various parties began to raise public questions about how much money was collected by CTIA from the industry corporations and what Carlo's SAG had done with it.

Within the companies that were paying for the research, the dispute gave those who were skeptics of the program from the outset a forum to air criticisms. Within the government agencies responsible for regulating wireless technology, primarily the FCC and the FDA, officials were unhappy with the delay that the dispute caused in the research effort. Within the trade media, journalists who were following the scientific effort covered the controversy, bringing new spark to their often routine dispatches. With funds cut off and future support uncertain, Ian Munro and Don McRee told Carlo they might leave the WTR program. At Carlo's urging they agreed to stay, as long as the money didn't dry up again.

The dispute threatened the viability of the program itself. For the next two years, the disagreement between Carlo and Wheeler grew increasingly bitter. At its worst, Wheeler accused Carlo of not adhering to the budget; Carlo accused Wheeler of spending dollars inside CTIA that had been earmarked for the SAG program. As Carlo pressed on with an ambitious and independent research agenda that was designed to determine the nature and severity of potential health risks—studies that might never provide public assurances that cell phones were safe—the CTIA withheld funds from Carlo's research program. When Carlo had to inform his scientific colleagues throughout the United States and the world that they had to suspend their research due to lack of funds, a number of them blamed his leadership rather than the industry.

Eventually, in 1997, a compromise was reached as Wheeler and Carlo signed a restrictive memorandum of understanding that significantly curtailed the scope of the scientific work that had originally been planned under the program. It was far from what Carlo wanted or thought should be done. But it did let some vital research go forward.

• • •

In the middle of his political-financial battles with Wheeler, Carlo also found himself in a bit of a political-scientific battle with his program's own Peer Review Board (PRB). On July 15, 1997, Dr. John D. Graham, director of the Harvard Center for Risk Analysis, and Susan W. Putnam, the project director who coordinated the peer-review efforts, sent Carlo a six-page letter of assessment. The letter started out in polite academic fashion by thanking Carlo for having participated in a recent meeting, then listed a number of the WTR's accomplishments. Then it got to its real point, in a section titled "Limitations." Among the key points was this one: "The first shortcoming of the WTR program stems from our perception that on occasion it has created expectations in the scientific, business, and regulatory communities that have not been matched by meaningful funding of original research. . . ."

Once again, Carlo found himself caught in a vise. This time he was being squeezed between the industry's limitations on funding and the scientific community's desire to do more research.

The letter from Graham and Putnam also criticized what they called "the rather ill-defined and limited relationship between the WTR and [Peer Review Board] PRB." Graham and Putnam recommended that the PRB they were running should have a far broader role in the WTR's decision-making, and not be relegated to a "limited, though not insignificant, scientific role."

And finally, they wrote: "We urge the WTR to document more publicly not only its accomplishments but also its spending for the first three years." Carlo had indeed created unnecessary doubts among the journalists who wrote about his research effort by refusing to make his financial books available for inspection by the media and the scientific community.

It was a real mistake on my part, from a public-relations point of view, to not provide copies of our audit statements to the media and others. We were rigorously audited on a regular basis—and the industry knew where every penny of the money was spent. I kept seeing it as a private matter between our research team and the industry, and didn't want to see it degenerate into a discussion of how much this scientist is getting as opposed to that scientist. But I failed to see that we were in effect a quasi-public effort. People were counting on us. They had a right to know that the money was being well spent.

A GOVERNMENT INTERVENTION

While Carlo and Wheeler were at war over the cash-flow problem, a government report caused them to be thrown together once again as uneasy allies.

At the request of Rep. Edward Markey, the General Accounting Office (GAO) had investigated the relationship of the cell phone industry and Carlo's research program. While the GAO investigators viewed favorably the scientific part of the program Carlo had set up, they concluded that it was not sufficiently independent of the CTIA. The GAO did approve of the separately funded Peer Review Board, in which Carlo placed funds earmarked for peer review in an escrow account, that the Harvard coordinators of the board then drew upon as needed. The GAO investigators saw this arrangement as more desirable than SAG's pay-as-you-go arrangement. The GAO wanted something akin to that to guarantee the independence of Carlo's research program. The GAO issued a draft report recommending that a firewall be established between the CTIA and Carlo's SAG program. Some time later this would lead to the creation of WTR, the successor to SAG.

Carlo and Wheeler met to discuss the GAO report that would be made public that spring. They agreed that the escrow fund approach was a good one, and they adopted it. A one-way deposit-only escrow fund was put in place to hold the money for the program. Carlo and Wheeler went one step further. They also agreed to establish an independent corporate entity to handle the business of the SAG.

"If you want to be independent, be independent," Wheeler told Carlo. The establishment of an independent corporate entity would serve them all well.

Thus, the Wireless Technology Research Limited Liability Company—the WTR—was established with the three members of the SAG, Drs. Carlo, Munro, and Guy, as its operating board, following GAO recommendations. The CTIA was a founding member but carried no formal role forward in terms of the scientific program or its management. Following the WTR Certificate of Formation upon which Wheeler and Carlo agreed, an Audit Committee was established to oversee an annual independent audit of the WTR's books and to specifically ensure that any payments to Carlo, personally, from the escrow fund were first approved by a member of the Audit Committee.

The focus on the formation of the WTR had masked the unsolved problem of the money, but only temporarily.

CTIA had continued to pay the invoices that Carlo's group had submitted through March 31, 1995, after they had been approved by Liz Maxfield. On June 1, a payment schedule was put into place by CTIA; they were to deposit $1,250,000 in the escrow fund quarterly—an amount that was one-quarter of the $5 million that CTIA was now calling WTR's annual research budget. Carlo disagreed with the designation of this amount as a budget, but was assured by Maxfield that this was a necessary designation for CTIA's internal accounting structure. She also assured him that the money that he needed would be there for the research. He was still operating on the assumption that his program would receive the approximately $10 million that he had budgeted for 1995.

The arrangement worked for eight months—until the WTR's cash-flow needs exceeded the funds that had been placed in escrow by the CTIA. WTR was operating on a $10 million budget and CTIA was assuming a $5 million budget for the research. As Carlo continued to ask for the money needed to cover the research, Wheeler began questioning Carlo's use of the funds, in particular the $3 million study Carlo had established to determine whether wireless telephones interfered with implanted cardiac pacemakers. While Wheeler had publicly celebrated the cell phone–pacemaker resolution as a CTIA triumph in his previous press statements, he now contended that the "fully independent" Carlo had gone beyond the scope of what his program was supposed to cover in doing the pacemaker research. Wheeler was now saying that the entire program was supposed to cover only cancer—a contention

Carlo believed was clearly inconsistent with the published research agenda and the other published descriptions of the program that had been in circulation for more than two years. Wheeler maintained that there was a financial shortfall because Carlo's program was focused not just on cancer but also on other health issues.

In a May 30, 1996, letter to Carlo, Wheeler wrote that the WTR had made its decisions about pacemaker research without first proposing a budget. The fact that they had the capacity to develop the needed science on the issue was good, but he felt that they undertook the project without a clear vision of how it would be paid for, which was not good.

Carlo's position was very different. In a July 5, 1996, letter to Wheeler, he responded with an analysis of the problem that had been prepared for the WTR board by Linda Solheim, the WTR's general counsel. Solheim had documented the decision making that led to the pacemaker study—including a request by the CTIA to have WTR involved. She also documented the fact that the pacemaker study budget item had been approved by the Joint Review Committee on January 19, 1996. And she noted that the CTIA's Arthur Prest had participated in the development of the pacemaker study protocol.

Within the industry, Wheeler used Carlo's independent work on pacemakers as an example of a public-relations problem that could be dealt with more efficiently with the new internal health and safety staff he had assembled within the CTIA.

• • •

In 1996 the WTR program was running a deficit because CTIA had stopped making deposits to the escrow fund to cover the budget. The tensions between Carlo and Wheeler were high. As more studies were approved for funding through the WTR's peer-review process, and as independent investigators supplied cost figures, it became clear that at least $42 million would be required to do the work that had already been identified as important and necessary under the WTR program.

Carlo and his team had prepared a thick black notebook that contained all of the planned research studies and the timetable for completing the work. Since Carlo and Wheeler really had very little to do with each other at that time, Carlo invited the CTIA's vice president for health and safety, Tom Lukish, to his office for a briefing. Carlo showed Lukish the budget notebook that listed $42 million in research, and explained that if they were also to do whole-life animal studies, in the event that the earlier studies produced positive findings, then the budget figure would rise to $60 million. Lukish said: "That doesn't bother me. I was in research and development at duPont. We felt we were doing good if we were within 20 to 25 percent of the projected budget." Lukish paused, waved his hand back and forth in front of his eyes, and added: "You never know with research." Lukish went on to explain the political realities of the budget, from Wheeler's viewpoint: "My boss has told me that we have a $5 million budget per year." But as he thought again about the research that needed to be done, Lukish, a graduate of the U.S. Naval Academy, surprised Carlo and his advisers by saying: "Look, we're not going to treat you this way. We're going to do what's right." The CTIA vice president picked up his copy of the WTR budget and left Carlo's townhouse office for the one-block walk to CTIA headquarters.

The next and last time Carlo talked with Lukish was a day or so afterward. Lukish telephoned Carlo and said: "I'm not going to be working here any more. I did what I could. Good luck." Lukish and Wheeler had come to a parting of the ways.

A RARE EXECUTIVE: PROMOTING RESEARCH

"I felt we had a responsibility to pursue the research wherever it led to," Lukish recalled, some four years after he and Wheeler had come to a parting of the ways. Lukish had come to the CTIA in 1995 after having retired early from a career that had taken him to a top position with the duPont Corporation, where he ran a department for safety, health, and environmental affairs. Lukish said he had planned to stay in that CTIA job for two years, but left after just one—for what he described as personal reasons. Lukish had a unique vantage point from which to observe the clashes of Carlo and Wheeler. He had seen it from the inside, as Wheeler's vice president in charge of health and safety—and overseeing Carlo's program. Lukish agreed to share his unique perspective in an interview with Martin Schram.

"George Carlo is a very, very honorable guy. He's principled. He's ethical and he's honest. . . . What George had created was a completely innovative new approach to researching . . . totally outside the influence of the industry. George had gotten into what I perceived as a normal research endeavor. It led down some blind paths, as research will. Many people in the industry needed to understand that. But Tom Wheeler did not. . . . This is what I tried to explain to Tom Wheeler: In my business experience . . . research budgets are very dissimilar to a construction project where you know most of the facts and there isn't any pioneering research or any invention that needs to take place. So a construction project budget should come in at plus-or-minus 10 percent. . . . A research budget is much more flexible—it could be plus or minus 25 or 30 or even 40 percent-depending upon the extent of the innovativeness and the pioneering research that is being attempted. And that's basically what George was into—he was creating new avenues of research. . . . If this was explained to the CEOs of the wireless corporations, they would understand that type of budgetary problem. Tom Wheeler does not. He did not grasp the fact that the research was pioneering and needed to be put into perspective. And we really could not expect, based on what George had shown us, that we could stay within that $25 million budget. . . . That was very, very difficult for Tom Wheeler to accept."

Did Lukish believe the industry wanted a program that might have shown problems?

"One of the reasons I came into CTIA was that I was very impressed with the business approach to this research. . . . the way the research was set up—to probe the existence of a risk—was exemplary. That's what impressed me. Here was an industry that had set up an independent research organization. And the industry had its hands off—at least when I was there it did not try to influence the research. . . . I don't know what the state of the research is now, but if they have let it deteriorate . . ." His voice trailed off, and he explained that he really didn't know anything about the recent state of the science. Research often does not result in consensus, he said, as scientists can be quite critical of each other's findings and defensive about their own. "I do know that it is difficult to get scientists to agree on a conclusion."

Lukish was asked to explain the dynamics of the personal and professional clashes of Tom Wheeler vs. George Carlo—clashes which had a profound affect on the information that would reach the public. "Tom Wheeler is a strange study in human nature," Lukish said. "He is not the type of business manager that I was comfortable working with. I think he is much more headstrong than George Carlo. . . . Tom Wheeler was not happy with what had gone on. He thought that George had mismanaged the budget over the last few years, didn't give him enough warning that this excess had occurred. He felt that he, Wheeler, was backed into a corner, that George could have given him a bit of a heads up, that sort of thing . . . But from my point of view, George being a scientist . . . they are more interested in the science, and pursuing the science, and the objectives of the science, than in the money. And that is what I tried to explain to Wheeler. And Wheeler was very, very critical that George had exceeded the budget by this amount. I tried to explain to Tom, well, okay, it's done. But you have to try to understand that George Carlo is a scientist. He's deeply involved in the research and trying to keep the research balanced. He's not an accountant or a business manger, he's a scientist—and you have to try to understand that. Well, Wheeler didn't want to understand that. He felt it was mismanagement, pure and simple.

"Basically what Carlo was doing was to try to invent a solution—I call it pioneering research. And the basic purpose of going down a path is that, if the research proves to be positive . . . it has to change the budget . . . in any highly technical research oriented corporation.

"I would trust George with anything. If George's position was that he needed to pursue something because it didn't look right, I would accept that."

• • •

As tensions continued to rise through the summer of 1996, attorneys for both Wheeler and Carlo stepped in and brokered the compromise that defined specifically what studies would be covered with the remaining funds the industry was willing to put into the program. And the attorneys almost had to step in again—literally—at one point during a meeting on the subject that took place in the conference room at Wheeler's office. At one point, Wheeler and Carlo were sitting side by side, with an overhead slide projector between them, when they got into an argument over a matter that would have seemed esoteric in other circumstances: the difference between budgeted funding for a project and a commitment to fund a project. To them it became a matter of parentage and manhood. Voices were raised, then they were shouting as they stood jaw to jaw, with only the projector between them as a referee.

Wheeler: "Are you calling me a liar?"

Carlo: "Well, if the shoe fits."

Top advisers and lawyers—would-be cornermen and cutmen—just stared. "I thought they were going to start punching," said Jeff Nesbit. But instead, they just went to their neutral corners—Wheeler to his office down the corridor, Carlo to his townhouse down the block. From then on, they dealt with each other at block's length, via intermediaries and attorneys. And it was just a matter of weeks before the deal was done. Wheeler and Carlo signed—separately—the Memo of Understanding that outlined the terms under which CTIA would continue funding the WTR research program.

According to the limitations of the agreement, a number of important research efforts already under way had to be scrapped for lack of funds. Among them: five major epidemiology studies, a program to develop methods to certify the amount of radiation from phones, a program to evaluate the impact of cellular phone base stations (towers and other relay fixtures), studies to examine the health impact of phones on implanted cardiac defibrillators, studies assessing the impact of wireless phone usage while driving, and 90-day exposure studies to assess the impact of wireless phone radiation on laboratory animals.

For Carlo, it was a disturbing litany of lost research opportunities. But of all the canceled programs, the human epidemiology projects were the most expensive—and perhaps most crucial. No new money was being made available by the industry to fund these five specific studies that had been approved by the Harvard-based PRB and the WTR's Epidemiology Working Group. The peer-approved studies that the industry would not fund were: two case-control studies of adult-onset leukemia, two that covered salivary gland tumors and brain cancer, and a community-based study of illness among 30,000 cellular phone users. In addition, the industry declined to provide money for ongoing surveillance of people who were using cellular phones and the health problems they might develop over time.

Carlo's WTR had already notified top scientists at schools including the Universities of Illinois, Southern California, Massachusetts, North Carolina, Texas, and Washington; State University of New York at Buffalo; and Boston University that their projects had been approved by the Peer Review Board, pending funding. Now he had to notify each of the scientists that there would be no money—and in a number of cases, the scientists made Carlo the prime target of their displeasure.

The PRB expressed its strong disappointment that funding had been cut for epidemiological research. The WTR's working group experts and the government's Interagency Working Group on Radio Frequency Radiation all emphasized how critical it was to continue with all of the research. But every time Carlo pressed Wheeler to fund these human health-effect studies, the CTIA president replied that the industry had committed to provide $25 million for research—and that was it.

"This is our financial commitment," Wheeler told Carlo in one meeting in his CTIA office. "You're independent. Do what you have to do to come up with the money you think you need." Carlo and Wheeler even talked about the possibility that the National Institutes of Health (NIH) might fund the epidemiological study. In 1996, Carlo did seek alternative funding sources, as he and epidemiologist Dr. Nancy Dreyer met with officials at NIH.

Another group Carlo approached was the Personal Communications Industry Association (PCIA), a trade association seen by CTIA as a competitor. Carlo met with PCIA president and CEO Jay Kitchen and talked about the need for tracking and monitoring of health problems among people who would be using the next generation of mobile phones—personal communication system phones (known as PCS systems), which are digital phones that can connect with the Internet and transmit data; they were just coming into use in 1996.

Kitchen arranged for Carlo to brief a few officials from the PCIA staff. They were interested enough to push the proposal to the next step—and soon Carlo was addressing the PCIA's board of directors. He handed each board member a written proposal for his proposed study: A group of PCS phone users would be assembled and entered into a database for biannual follow-up, looking at death information and the incidence of cancer. Warranty cards accompanying each phone would include an informed consent to participate in a follow-up study—thus getting around the legal barriers and privacy issues created by the Busse lawsuit in Illinois. The customer information on those cards would be the content of the database.

Carlo's proposal to the board was straightforward and low-keyed. But much to his surprise, it proved to be rather provocative.

After I addressed the PCIA board, the meeting became contentious. The members of the board got into a heated discussion about whether the post-market surveillance was the responsibility of the service providers or the manufacturers of the phones. It soon became the carriers versus the manufacturers—and I just looked on, as a bystander. No decisions were made.

Two weeks after that meeting with the PCIA board, I was summoned to Wheeler's office. When I arrived, he handed me a copy of my own proposal that had been distributed at the PCIA meeting. He looked to see if I was surprised that he had it. I was not. In no uncertain terms, he told me: "You work for CTIA. You work for me—not PCIA."

Meanwhile, the PCIA board decided not to pursue the matter. "Our board just wasn't interested in going ahead with George's proposal," PCIA president Jay Kitchen recalled. Kitchen added, when asked, that Tom Wheeler had never called him about Carlo's proposal. "I never discussed the issue with Tom at all," he said. ". . . Basically, we just took a pass on it."

• • •

All of the science experts agreed upon the importance of ongoing epidemiological studies and post-market surveillance, which would include setting up a database for all complaints that are received from cell phone users with health problems. The FDA's Dr. Elizabeth

Jacobson made this clear to the industry when she appeared before the CTIA board of directors in 1997. She told the executives that post-market surveillance should be considered an ongoing cost of doing business.

Later that year, I met in my conference room with several top Motorola executives—including John Welsh, Chuck Eger, and Norm Sandler—to review my plan for ongoing postmarket surveillance. After I reviewed the entire plan with them, John Welsh gave a sharp but unmistakable reply. He held up his right hand, formed a circle with his thumb and index finger, and said: "Zero. Zero surveillance. We're going to do enough research so that we can prove safety - and then we can stop doing research." A week or so later, I gave the same presentation at CTIA headquarters to Jo-Anne Basile and several others, including Eger and Sandler. Basile said that they would not fund my post-market surveillance model, saying it was not appropriate for cell phones because cell phones, after all were not pharmaceutical drugs. When CTIA took the side of Motorola in the debate, I knew that I had been used. Our program was supposed to lay the groundwork for post-market surveillance. Now they were all walking away from that.

VERY POLITICAL SCIENCE—I

December 3, 1998: The CTIA board meeting had already been underway for some time when George Carlo walked into the plush banquet room of the ornate Sheraton Carleton Hotel, only three blocks up 16th Street NW from the White House, just when Tom Wheeler's staff had instructed him to show up. About 75 cell phone industry executives and a few invited guests were sitting at tables that were arranged into a big horseshoe. Tom Wheeler was sitting at the base of the horseshoe, and an executive Carlo did not recognize was speaking.

Carlo thought to himself about how times and circumstances had changed. He didn't feel much like an insider at all—and for good reason. He had clearly been tagged as an outsider by those within. In the early days of the SAG, nearly six years ago, Carlo had always been invited to participate in all of the board meetings from beginning to end. He had always been encouraged to stay and participate throughout the day, have breakfast, and be in the mix of science and small talk with the CEOs and chairmen of the big telecommunication companies. Indeed, in years past, he had helped plan the program and develop the agenda.

I had been a trusted part of Wheeler's inner circle. But those days were over. Wheeler had already decided that I had outlived my usefulness to him, and he was now trying to move me out without any more controversy. Since the flaps over the funding and pacemaker interference in 1996, I had become a thorn in his side. For sure, Tom's and my personal relationship had all but disintegrated. The talk about buying sailboats together and vacationing together with our families was a distant memory. We exchanged greetings, but that was all.

Today, I had been asked to give what was being called a "report on the program." I was not privy to the rest of the agenda. I was allowed into the room only five minutes before my presentation. Jo-Anne Basile, now Wheeler's person on the health-effects beat, met me outside the room with curt directions about what I was to do once inside. Jo-Anne and I had enjoyed a difficult working relationship since she took over the health and safety program after Tom Lukish had left the CTIA.

Jo-Anne's instructions to me went something like this: Tom will call you to the podium. Give your presentation. Ten minutes tops. Tom will handle the questions and answers and call on you when you are needed. There is a lot on the agenda and your briefing is a small part of it.

"Thanks, Jo-Anne. Do they have the materials I sent over last week as handouts?"

"Everything is in their briefing books." Don't bother to bring your slides, I had been told.

Carlo spoke to the CTIA board for just ten minutes, reporting that the WTR's study of all of the known science in the world,

Dr. Bill Guy – the Scientific Advisory Group's mentor and dean of research in the area of radio wave physics.

Dr. Arthur (Bill) Guy (left), Dr. Ian Munro (center) and Dr. George Carlo (right) during a Scientific Advisory Group board meeting in 1994.

Dr. George Carlo (left), Dr. Ian Munro (center), and Dr. Bill Guy on the Chesapeake Bay during 1994 fishing trip with CTIA staff.

Cellular Telecommunications Industry Association head Tom Wheeler (left, in baseball cap) with Dr. Jim Tozzi, Scientific Advisory Group's liaison with the U.S. EPA.

ALL ILLUSTRATIONS COURTESY OF DR. GEORGE CARLO, UNLESS OTHERWISE NOTED.

Jeffrey Silva, Washington-based correspondent for the trade publication *Radio Communications Reports*; he was often critical of Carlo and his research program in its early years, but then reported extensively on the health risk findings.
CREDIT: *RADIO COMMUNICATIONS REPORTS*

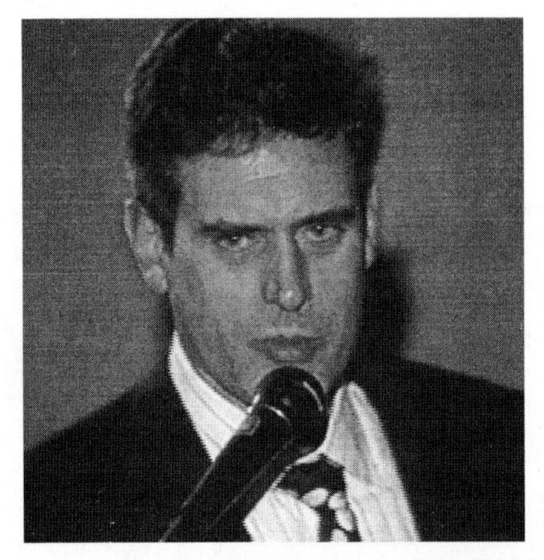

Dr. Kenneth Rothman of Boston University who headed the study of deaths among 250,000 cellular phone users.

Mr. Thomas Lukish, CTIA Vice President for Health and Safety who left CTIA in the midst of the flap over WTR funding.

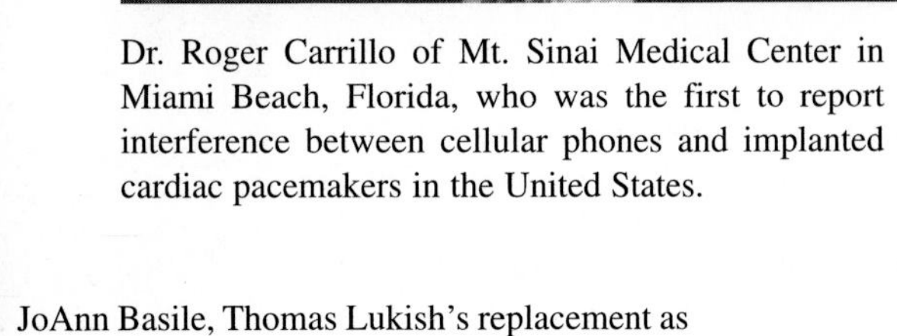

Dr. Roger Carrillo of Mt. Sinai Medical Center in Miami Beach, Florida, who was the first to report interference between cellular phones and implanted cardiac pacemakers in the United States.

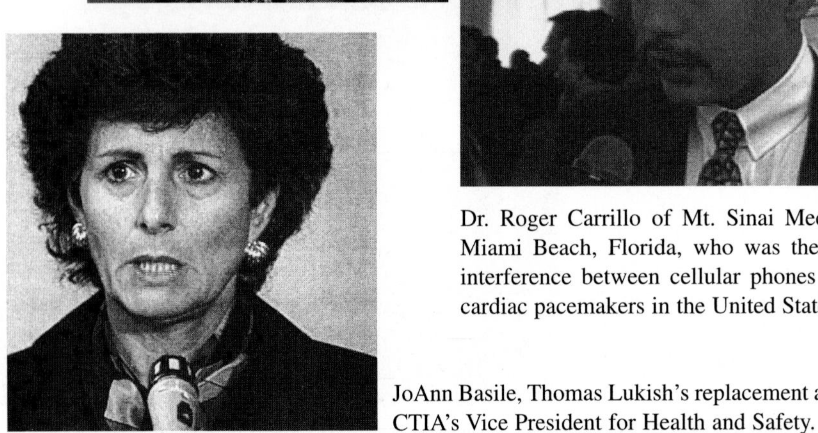

JoAnn Basile, Thomas Lukish's replacement as CTIA's Vice President for Health and Safety.

Dr. Om Gandhi of the University of Utah who was the first to show that cellular phone radiation penetrated deeper into the heads of children than adults.

Dr. George Carlo (left) demonstrating the use of a wireless phone headset to Brian Ross, Chief Investigative Correspondent for ABC News' 20/20, in segment first aired in October 1999.
CREDIT: STAN HANKIN

Dr. Susan Putnam, of the Harvard Center for Risk Analysis, who coordinated the work of the Peer Review Board.

Dr. George Carlo (at the podium) and JoAnn Basile (standing with back to camera), CTIA Vice President for Health and Safety, during a contentious exchange at the Second State of the Science Colloquium in Long Beach, California, in June 1999. Basile told Carlo, "You have caused us a few sleepless nights. . . ."
CREDIT: STAN HANKIN

Wave guide system showing rats are unrestrained and thus their whole bodies are exposed to radiation, not just their heads. This design was rejected by Carlo's researchers as inadequate for cellular phone studies.

Two antennae that were developed with Stealth Bomber technology to simulate human exposure to radio waves.

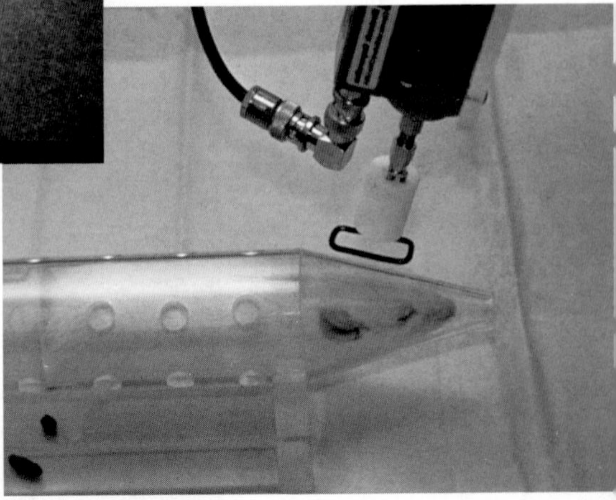

The head-only exposure system Carlo's team set up that shows rat in tube and antenna attached to coaxial cable that goes to cellular telephone. This was a scientific breakthrough many experts thought could not be achieved.

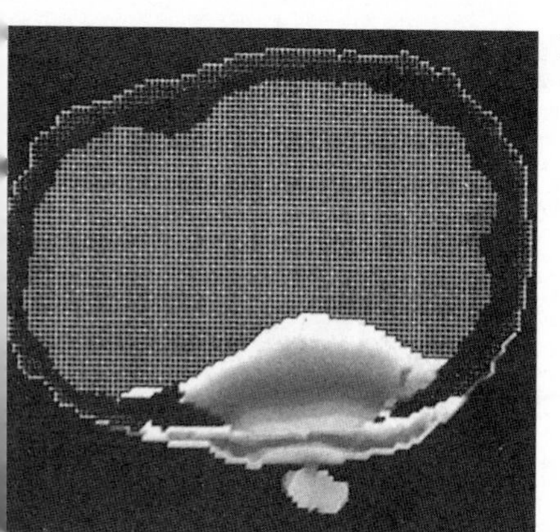

Radiation penetration in head of adult;

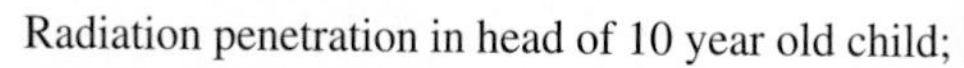

Radiation penetration in head of 10 year old child;

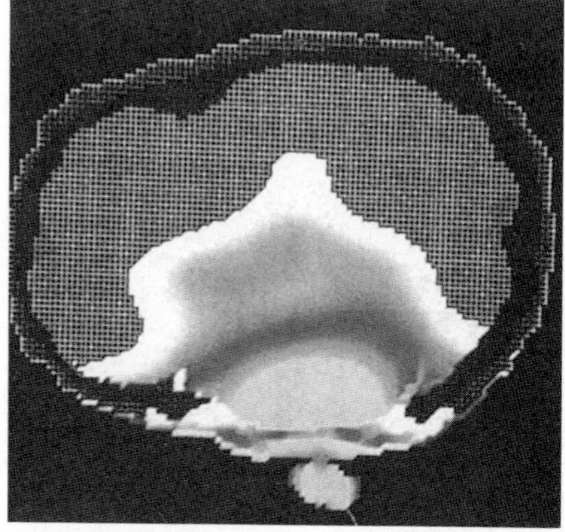

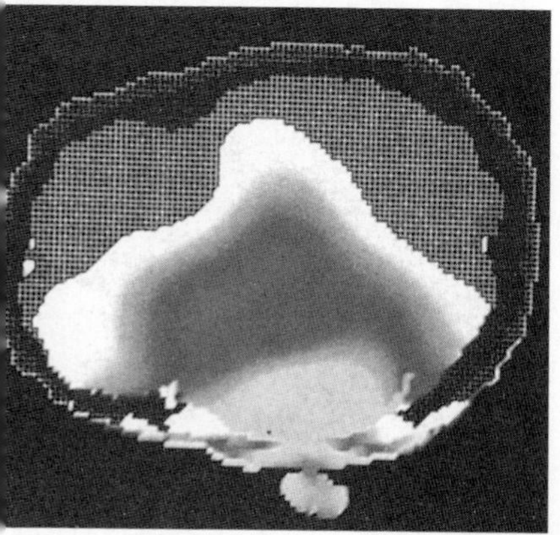

Radiation penetration in head of 5 year old child;

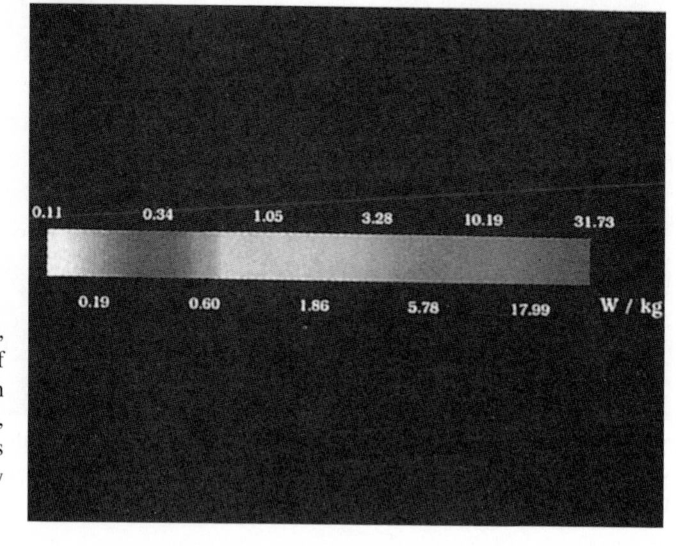

From the work of Dr. Om Gandhi, a computer-image comparison of the heads of adults and children (all adjusted to the same scale), showing that radiation penetrates younger skulls far more deeply than those of adults.

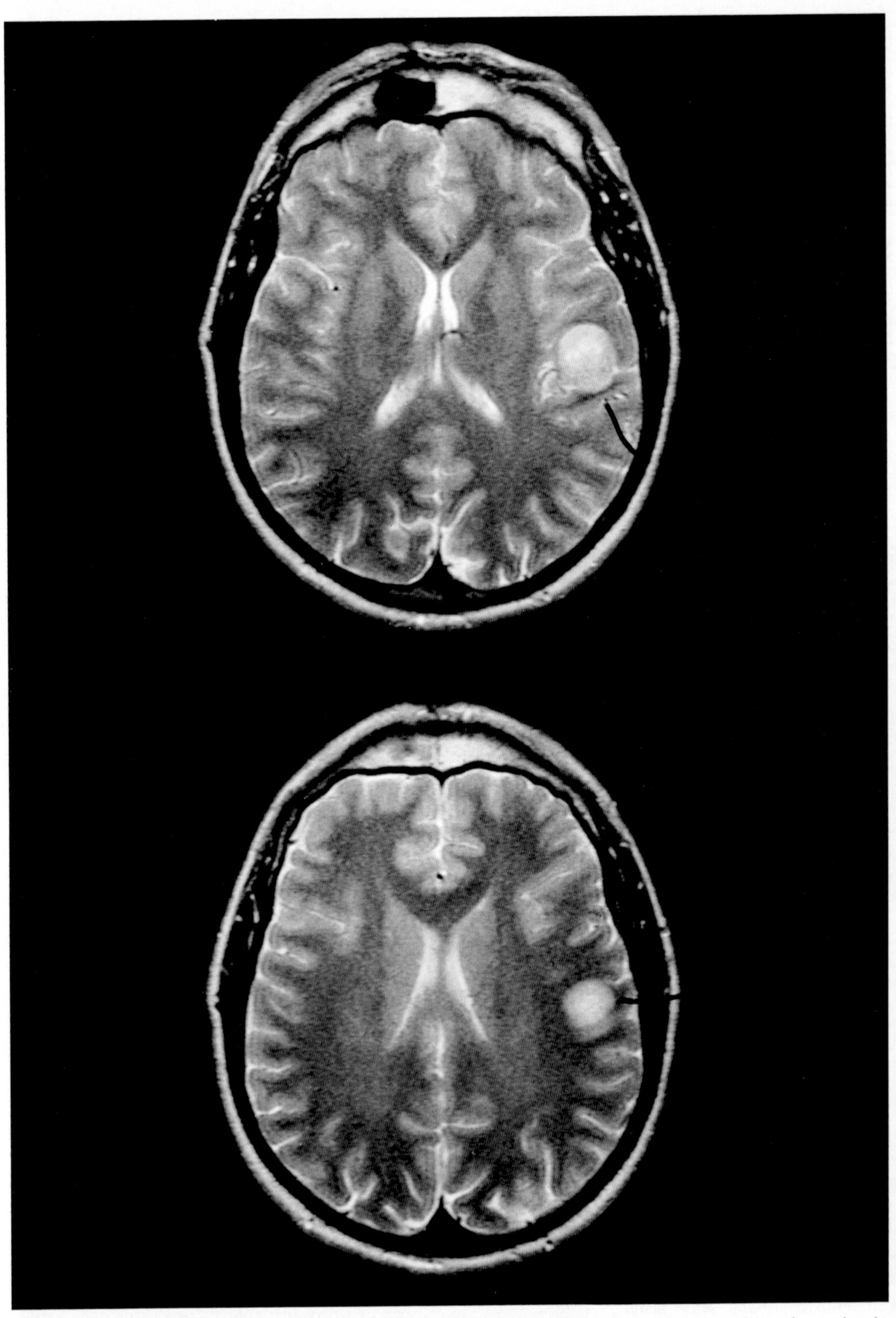

Two X-rays of a cellular phone user show a brain tumor located in the area next to where the patient's phone antenna was held during calls. The cancerous mass is visible on right side of head, indicated by the black line.

Graphic showing four different ways that DNA can be damaged—one of which involves actual breakage of DNA strands.

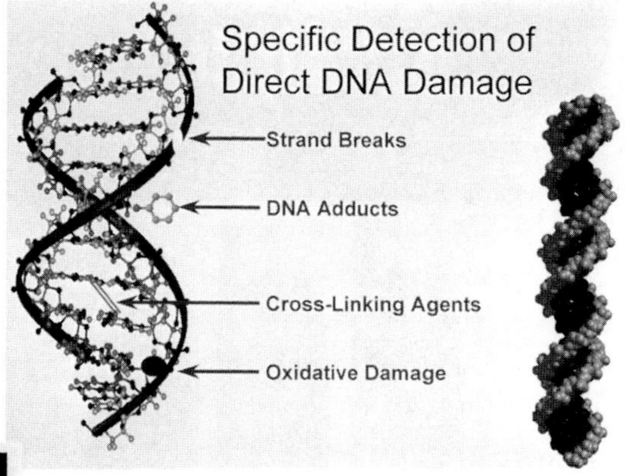

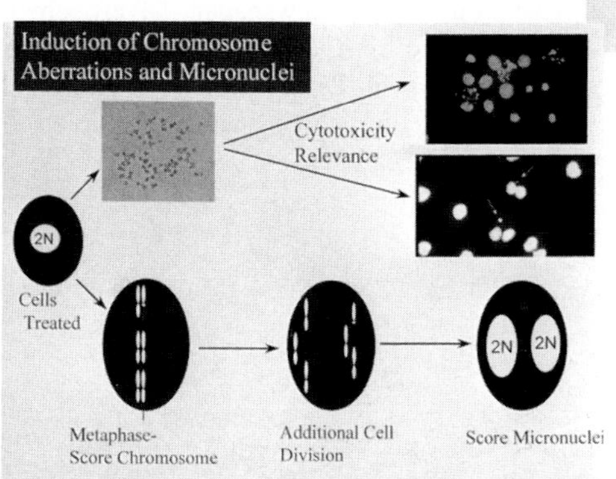

Graphic showing normal blood cells (top box) and blood cells that have been genetically damaged and have developed micronuclei (lower box).

Photo of a pacemaker prior to being implanted in a patient. In order to make the device immune to cell phone interference, the dual wires extending from the upper left side of the pacemaker need to be fitted with special filters. With this recommendation Carlo's group solved the problem of interference from wireless phones.

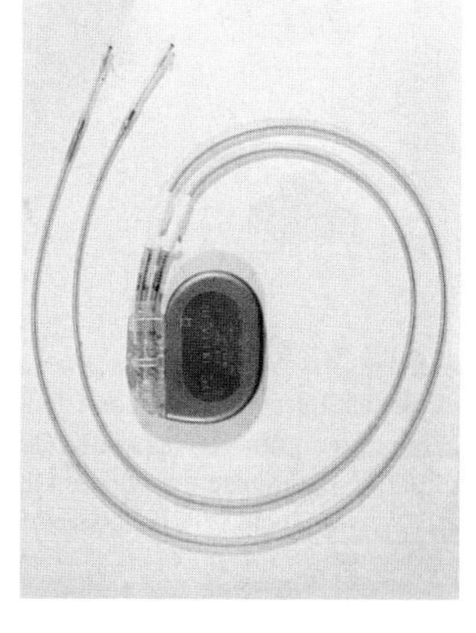

MN-BN CELLS: FOLD INCREASE OVER CONTROLS
8.0
6.0
4.0
2.0
0.0
1 W/kg
5 W/kg
10 W/kg
ANALOG 1 ANALOG 2 TDMA 1 TDMA 2 CDMA 1 CDMA 2 PCS 1 PCS 2
TECHNOLOGY

Graph showing data from the studies done by Drs. Ray Tice and Graham Hook for Carlo's research effort. For every type of wireless phone tested, the formation of micronuclei increased as the amount of radiation from the wireless phone increased. The graphic also shows that the number of blood cells that have micronuclei doubled when the cells were exposed to radio waves at 1 watt per kilogram SAR. That level is actually below the FCC's so-called safety guideline of 1.6.

Full page newspaper advertisement placed by AT&T showing phones decorated with Disney cartoon characters: a marketing strategy aimed at parents of young children.

Thomas E. Wheeler, President of the Cellular Telecommunications Industry Association; Wheeler appointed George Carlo to coordinate the wireless industry's research on cell phone safety.
CREDIT: *RADIO COMMUNICATIONS REPORTS*

including studies commissioned by the WTR, had basically shown no proven health problems for cell phone users. Several widely reported studies by scientists outside the WTR program had shown some positive findings of biological effects from cell phone radiation. But they all had flaws of one sort or another. Dr. Henry Lai's DNA breaks, Dr. Leif Salford's leakage in the blood brain barrier, Dr. Ross Adey's protective effect on cancer, Dr. Michael Repacholi's increased risk of lymphoma in mice, and Dr. Mild's increased risk of headaches—all had scientific question marks that had led the WTR and peer-review experts to believe the findings that indicated possible health risks were probably chance occurrences and were not to be considered as valid indications that users of wireless phones were putting their health at risk

Meanwhile, it was also true that the shortcomings in scientific procedures used in those earlier alarming studies, which found risks, also existed in many of the other early studies that showed no harm or health risk—studies which had been hailed by the industry as providing a clean bill of health for its products. If both types of studies were flawed, what should be the conclusion about health risks? The work of the WTR had indicated it was quite possible that those reassuring studies could also be no less valid than the troubling ones. But Carlo did not get into that point at this board meeting. Instead, after his ten-minute report he stood at the podium alongside Wheeler and, for about five minutes, responded to general questions from the board.

I was caught off guard by Tom's next move. He turned to face me as we stood side by side at the base of the horseshoe of tables and, while still speaking to the entire board, he addressed his next words to me.

"For the past five years, you have done a super job running this program. None of it would have been possible without you. We all owe you a debt of gratitude. Thank you."

He then began applauding, alone at first, but as he coaxed the rest to join in, they did.

I was shocked, but managed a "Thank you, Tom."

Then I left, accompanied by Jeff Nesbit, our WTR liaison with the

industry and government. As we crossed 16th Street for the four-block walk to my office on N Street, I asked him: "What the hell was that thank you and applause all about?"

"That, George, was a Washington-style kiss-off," Jeff told me. "This was the last time Wheeler expects you to speak to his board. You are done—and my guess is that Wheeler's view is 'good riddance.'"

"Like this has been a picnic for me?" I said, which caused Jeff to laugh. "Let's finish this program and move on. I have had enough of cellular phones, CTIA, and Tom Wheeler and his minions."

Jeff said, "Consider this a success. The industry has gotten the results they had hoped they would get—and you delivered them. That's worth a lot in this town."

PART THREE

CHAPTER TEN

FOLLOW-THE-SCIENCE: *RED FLAGS*

AS OF THE FIRST WEEK of December 1998, the state of the world's known science on cell phones and health risks basically came down to this: No proof of risk, no proof of safety. What the science was telling users of wireless phones was: It was unlikely that their mobile phones would cause health problems—which was just what Carlo reported to the CTIA board on December 3, 1998.

But just 18 days later, Carlo saw his first serious red-flag warning: A new Wireless Technology Research (WTR)-sponsored study in North Carolina—done with the new exposure system developed by Carlo's team—had produced an alarming finding: Cell phone radiation did indeed cause genetic changes in human blood cells. This would prove to be just the first in a series of new red flags Carlo would receive in the weeks to come. Together, these red-flag studies would become the basis for what he quickly saw as a new public-health imperative. The new findings meant it was clearly necessary to look back and re-evaluate all of the science that had gone before. It was also imperative that a major new round of research must begin.

Neither the science, nor the politics, of cell phone safety would ever be the same.

GENETIC DAMAGE IN HUMAN BLOOD

Micronuclei—First Findings:

December 21, 1998: "Your in-box is overflowing. I'll keep you clear for the next couple of hours if you promise to stay off of the phone and go through your mail." Lisa Joson, Carlo's assistant, had always tried to keep him focused, a noble and necessary effort. But sometimes it just wasn't possible. And so, that day, Carlo had put in several focused minutes of in-box management when the door to his office opened and Marjan Najafi, who was coordinating the WTR's toxicology program, peeked around the doorjamb and asked: "Do you have a minute?"

Carlo looked up. "Can it wait, Marjan? I'm swamped and Lisa is cracking the whip."

"I think you will want to see this—now."

Najafi handed Carlo her own short summary attached to a report that had come in the night before. It was preliminary results from a series of DNA damage studies being done by Drs. Ray Tice and Graham Hook of the Integrated Laboratory Systems in Research Triangle Park, North Carolina. The project was being coordinated by Don McRee under a contract with the WTR. (Najafi's role was to be the first to review all study findings. She would prepare a summary of the findings, then pass the summary and the report to Carlo for a final internal review. The report would then be forwarded to Dr. Susan Putnam at the Harvard Center for Risk Analysis, who would coordinate the outside peer review.)

The studies were of special interest because, while not part of the WTR's original research plan, they had been added to help interpret the controversial 1994 comet assay findings of Drs. Henry Lai and N. P. Singh which suggested that cell phone radiation causes DNA breaks. Indeed, the genesis of the Tice and Hook study illustrates the degree to which outside peer-review efforts were utilized to monitor every step of the WTR process. And since the findings would prove to be so controversial, with the industry seeking to dismiss and discredit the research, it is important to note the extent to which Carlo had insisted the entire process be peer-reviewed by independent experts—from inception to conclusion. After seeing Dr. Lai's findings, Carlo had brought together three groups of world-renowned scientists to advise on what sort of additional studies might be needed. The Harvard-based Peer Review Board (PRB), Dr. Ian Munro's Toxicology Working Group, and the WTR's Genotoxicity Working Group came back with a recommendation for five different types of new tests, one of which was a test to determine whether radiation increased micronuclei development. A request for proposals brought back more than 25 research teams around the world seeking to do the work. The proposals were then submitted to the PRB, which narrowed it down to two finalists. Tice and Hook got the job. They prepared study protocols that were then forwarded to Carlo, who sent them on to the PRB. After suggestions and modifications, the research began. Tice and Hook had looked at the induction of DNA and chromosome damage in human white blood cells. They did this by using a technique called the "micronucleus assay"—a test of blood cells exposed to cell phone radiation, which sought to determine whether the exposure caused the nucleus of the cells to divide into a number of micronuclei. The scientists measured damaged DNA by looking for fragments of chromosomes that formed membranes around themselves and appeared under a microscope as additional nuclei in blood cells. Then they prepared a quick summary and faxed it to the WTR.

Carlo glanced at Najafi's summary of the report—then stared at the paper and read it again.

I was stunned. Tice and Hook had run the experiments using analog signals in the cellular frequency band, and their preliminary results showed that following exposure there was a nearly 300-percent increase in the incidence of genetic damage in those human blood cells.

I looked back at Marjan and asked, "Are you sure about this?"

"This is what Dr. McRee has sent," she answered.

I called Don McRee, who was at the lab running the other tests that were part of the original plan. I left him a message to call back as soon as possible, which he did about noon.

"George, I thought you would be calling."

"Well, Don, what does this mean?"

He told me, "It means we have some genetic damage from the phone's antenna. It is only there, however, after the cells are exposed for 24 hours. During the three-hour exposure it doesn't show up."

"Why 24 hours and not three hours?"

"Don't really know, but we sure need to find out."

I agreed with him. Our protocols had called for a complete battery of genetic damage studies, including separate studies using bacteria, mouse lymphoma cells, and human lymphocytes. The results were in from those other studies, and they were all negative—no evidence of genetic damage at all. But this new finding of genetic damage, although preliminary, seemed to contradict the other results.

"Is this just an artifact?" I asked Don. "The other studies are all negative."

He said it was hard to tell, but the micronuclei studies were looking at something different.

This was why our Toxicology Working Group had recommended that we do the full battery of validated assays. It was not unusual for a positive effect to show up with one type of study and not with another. When the results were all in agreement, it was a straightforward conclusion—either positive or negative for genetic damage. But when you had what we now seemed to have—conflicting results—a more careful interpretation was necessary. Inevitably, more tests were in order.

"Is there anything obvious in the testing methodology that might explain it?" I asked.

He said there were no obvious flaws. The tests were run properly, with GLP—shorthand for Good Laboratory Practices—and good quality control. But Don said we needed to make absolutely sure the micronuclei that had developed were formed due to the exposure to radiation from the cell phone, and not due to heating in the test tubes.

Other experiments by other scientists in past years had been discounted because of the belief that it was a buildup of heat, and not radiation, that had caused results that seemed to show health risks. That was why we had invested millions of dollars in developing test systems that controlled for heating. But Don and I both felt we had to check once again—something might have gone wrong in the experiments themselves, although he didn't believe that anything had. He also thought we should add another dose of specific absorption rate (SAR) exposure level to the experiment, to see if a change in the SAR dose level would create a different response.

"OK," I said, "let's re-do it."

Carlo went back to Drs. Tice and Hook with the additions to the protocol. They all spent a good part of the Christmas season of 1998 in the laboratory, recognizing the urgency of the study. The question of genetic damage was critically important to understanding any mobile phone health effects. Without some proof of genetic damage, it was unlikely that any scientific experts would conclude the phones could cause brain cancer. With evidence of genetic damage, focused follow-up studies would definitely be necessary; with a finding of genetic damage, all sorts of health effects were possible.

Very few genetic damage studies had been done that were relevant to people using cellular phones. And while a number of early studies had indeed shown no genetic effects, Carlo and his team of experts had grown to doubt the validity of the *in vitro* exposure systems under which they had been done. These new studies by the WTR lab in North Carolina was the first to be operating with the new exposure system that had been specifically designed for this purpose—and its initial findings were different from those others that had shown no health effects.

We had cast our lot with the studies we were doing with the new exposure system. The WTR studies were intended to fix the problems that were identified in the older studies, and to be the new state of the art. We had made it a point of emphasis to make sure that shortcomings in the earlier studies were not repeated in the WTR studies.

We had to get it right.

• • •

"George, it looks like Don and his team have reproduced the genetic damage results." It was January 28, 1999, and once again, Marjan Najafi was the bearer of stunning news: The earlier indication of genetic damage in blood cells that were exposed to cell phone radiation had been replicated in the new round of experiments. Najafi handed me a set of tables and graphs that McRee had just sent from their lab in North Carolina.

I was stunned. Much to my surprise, the genetic damage had been confirmed.

Micronuclei – A Diagnostic Marker for Cancer Experts

The relationship between the presence of micronuclei and cancer is so strong that doctors from around the world are using tests for the presence of micronuclei to identify patients who are likely to develop cancer—so they can be properly treated. Indeed, after the 1986 nuclear disaster at Chernobyl, in the former Soviet republic of the Ukraine, international teams of experts who arrived on the scene used micronuclei testing as a vital tool for diagnosing cancer risks and saving lives.

All tumors and all cancers are the result of genetic damage, and most often that damage includes the formation of micronuclei.

In a critically significant review paper published in the August 2000 issue of the prestigious *Journal of the National Cancer Institute,* Dr. Marianne Berwick of the Memorial Sloan-Kettering Cancer Center and Dr. Paolo Vineis of the Unit of Cancer Epidemiology, University of Toronto, noted that the presence of micronuclei in living cells indicates that the cells can no longer properly repair broken DNA—and that this deficiency will "likely lead to the development of cancer." This means that the presence of micronuclei is now considered by expert scientists as an indicator of an increased risk of developing cancer.

In a 1996 review article published in *Mutation Research*, the primary review journal in genetic toxicology, Dr. J.D. Tucker of the Lawrence Livermore National Laboratory and Dr. R. Julian Preston, of the Chemical Industry Institute of Toxicology and an expert adviser on the WTR's toxicology working group, cited micronucleus studies as one of the most useful tests for cancer and genetic risk assessment. Over the past four years, groups of scientists in France, Germany, India, Italy, Sweden, Turkey as well as the United States have all published reports of their work that relies on the presence of micronuclei in blood as a clear indicator for the risk of cancer and other health hazards.

The most notable use of micronuclei testing as a cancer predictor occurred after Chernobyl, when scientists from Italy and Germany tested children who lived in the vicinity of the plant by taking measurements of the micronuclei found in their blood cells, identifying the children most at risk of developing cancer from the radioactive fallout. Once those children were identified, doctors began various preventative treatments to minimize the chances that they would develop cancer.

Scientists from the University of Pisa in Italy published their findings on micronuclei in *Mutation Research* in 1999; scientists from the Clinical University in Essen, Germany published their studies in the *International Journal of Radiation Biology* in 1996. Scientists at the Department of Toxicology, Faculty of Pharmacy, Gazi University Hipodrom, Ankara, Turkey have reported on their work (published in *Mutation Research* in 1998) using micronucleus tests as an indicator of increased cancer risks among teenage workers exposed to chemicals in engine repair workshops.

THE SPECIFICS OF THE WTR STUDIES

The Objective:

Two studies were conducted simultaneously to determine whether DNA damage was occurring in human blood cells that were exposed to mobile phone radiation—and if so, how it was occurring.

To do this, researchers first had to see if actual breakage occurred in the DNA, as Drs. Lai and Singh had reported in 1994. Then the researchers had to attempt a far more detailed test of the human blood cells to see if there had been formation of micronuclei—where cells would have a number of little nuclei, instead of just one normal nucleus. This would show that there was genetic damage, which could indicate a serious health risk, even if there was nothing quite as serious as actual breakage of the DNA.

The Procedure:

> In their tests, Drs. Tice and Hook made use of six innovations to correct for shortcomings in earlier studies by other scientists.

1. For the first time, the researchers were able to assure that the radiation was spread evenly throughout the test material in each tube. They did this by applying the computer model findings of Dr. Guy and only using the bottom portions of the tubes.

2. For the first time, the researchers were able to control for unwanted buildup of heat in the test tubes that could artificially skew the findings. They used internal fans and a circulating water bath to maintain a temperature of 37 degrees Celsius inside the exposure chamber that held six test tubes. They used two fiber-optic thermometers inside each test tube to assure that temperature was maintained at the center and sidewalls of the bottom of the tube that contained the human blood.

3. For the first time, the researchers were able to use a range of radiation dosage that went even lower than the Federal Communications Commission's so-called "safety" guideline of 1.6 watts per kilogram. The Specific Absorption Rate levels in the tests of Tice and Hook were at 1.0, 2.5, 5.0, and 10.0.

4. For the first time, the researchers used real mobile phones to generate the radiation to which test tubes would be exposed (instead of using commercial generators to simulate the phone signals, as had been done in past experiments).

5. For the first time, researchers used an actual human voice, recorded on a compact disc that was played through the phones. That provided a realistic exposure, because many scientists had long believed that changes in voice modulation produced changes in signal characteristics that could alter the radiation emissions. While past experiments had merely exposed blood cells to radiation by beaming a constant wave through signal generators, Carlo suggested the innovation of using an actual recorded human voice to talk to the blood in this experiment.

6. For the first time, the researchers used all types of mobile phones in a single experiment, in order to compare the effects of different types of exposure. They used analog, digital, and personal communication system instruments that can transmit data as well as voice.

The Findings:

> The "comet" test showed no evidence of DNA breaks in white blood cells—but the micronucleus test showed that damage within chromosomes did occur.
>
> There was no breakage in DNA when the blood was exposed for durations of just three hours or for a 24-hour period. Also, there was no chromosome damage found in blood that was exposed for just three hours. But when the blood was exposed to radiation for 24 hours, chromosomes were found to be damaged. This damage was indicated by the formation of micronuclei in blood cells (which normally have just a single nucleus).
>
> This damage was found when cells were exposed to signals from all types of phones—analog, digital, and PCS—at SAR exposure levels of 5.0 and 10 watts per kilogram. This damage showed that there were four times as many cells with micronuclei in

human blood that had been exposed to radiation as exists in blood that is not exposed to radiation. That is a huge change that is considered a significant effect.

In addition, chromosome damage from signals of two types of phones (a cellular digital phone and a PCS phone) also occurred at an SAR level of just 1.0 watts per kilogram—a level that is below the government's so-called "safety" guideline of 1.6. That damage showed there were twice as many cells with micronuclei in human blood that was exposed to radiation, compared to blood that is not exposed to radiation. That increase, while real, was not large enough to be labeled statistically significant.

The Significance:

For the first time we had conclusive evidence that mobile phones cause real genetic damage. This finding was so surprising that the first time I was informed of it, I thought it must be a fluke. So we went back, redid the experiments—and got the same results. Then we did it one more time—and got the same results again. The significance of the findings, for cell phone users everywhere, was unmistakable: it meant that their phones could be dangerous to their health. The significance of the findings, for the cell phone industry officials, was that they could no longer accurately say there was no evidence of harm from cellular phones—their standard assurance that cell phones were safe.

Beyond all of that there was one additional finding that struck me as more than just surprising—I thought it was alarming and I thought the government officials who set the standards would too. For the first time, we had found a real increase in genetic damage at a 1.0 radiation level that was below the FCC's 1.6 SAR level that is called the "safety guideline" by officials of industry and government.

To appreciate the significance of this finding it is important to focus on just what that FCC guideline really is and how and why it was established. The FCC's radiation emission guideline was established based on an assumption that there were no adverse effects at all below a radiation level of 40 watts per kilogram—and to protect the public, there must be a margin of safety that is 25 times lower than the known danger level where health effects had originally been found. So the FCC experts simply divided 40 by 25 and came up with 1.6 as the so-called safety guideline.

But now, the WTR had found significant genetic damage at 5 and 10 watts per kilogram—and some damage at levels as low as 1.0. Based merely on the WTR's findings of genetic damage at SAR levels of 5 and 10, if the FCC still believes there must be a 25-fold margin of safety to protect the public, it must recalculate its guideline. When the FCC experts divide 5 by 25, and 10 by 25, they will find that their new "safety" guideline, according to the FCC's own policy, must be between 0.2 and 0.4—figures far below the government's accepted 1.6 standard!

What this means for cell phone users is that the phones they are holding against their heads may well be operating in a dangerous range.

• • •

These were not just another batch of studies. These studies were the result of a six-year process in which Carlo and his highly respected science advisers had insisted on rigorous procedures that would assure proper methods of exposure, measurement, and verification. The studies were the collective efforts of some of the world's best scientists. The fact that they had not rushed into new research that would repeat old problems had been roundly criticized by some in the media, the government, and the industry. These studies, then, were a significant vindication for the WTR approach institutionally, and for Carlo, personally.

The deliberative scientific process had been conducted openly, encouraging the involvement of experts throughout the scientific world. For four years at the prestigious Bioelectromagnetics Society meetings, these methods were discussed in an open and active scientific exchange. The methods had been published and peer-reviewed in journals. Some $8 million was invested over four years in just the experimental designs. The studies were done and then replicated and expanded over a 19-month period. And as soon as Carlo got the report from Tice and Hook, he simultaneously sent it the Harvard for peer review, a process by which the findings and study procedures were sent out for review and recommendations to several prominent scientists whose identity would not be known to the researchers. Carlo also sent them to the Toxicology Working Group for additional peer input.

And because this round of WTR studies raised a red-flag warning about the safety of a very popular product, Carlo sent out one additional alert.

I immediately placed a phone call to Tom Wheeler.

"Tom, we need to meet. We have a problem here."

"What problem?" was his response.

"It looks like we have some genetic damage to deal with."

VERY POLITICAL SCIENCE—II

Before CTIA's top officials would meet with Carlo to hear about this new data, they wanted to have another science consultant help them deal with the data. Apparently, Carlo's status had slipped to the point where Tom Wheeler now wanted an intermediary to filter Carlo's interpretations of the research. To interpret Carlo's interpretation, CTIA chose Dr. Martin Meltz of the University of Texas. Tensions had grown between Meltz and the WTR ever since Dr. Guy's petri-dish studies had suggested that Meltz's earlier research may have been done with a flawed exposure system.

CTIA vice president Jo-Anne Basile followed up on Carlo's initial call to Wheeler by asking Carlo to send the preliminary reports from the North Carolina lab so that the industry's chief lobbyist and his aides could review them prior to the meeting.

"No, Jo-Anne, I can't do that."

"What do you mean, 'No'?"

"Jo-Anne, it is not appropriate for me to copy the results and send them to you prior to peer review," I replied.

"We have a right to see the data." She was adamant.

I did not agree with her, at least not before the peer review was complete, though I saw value in the industry being advised as soon as possible of the situation. This was not comfortable for me either.

I responded with a compromise. "If Marty Meltz is willing to be an outside peer reviewer of the data, that would be acceptable. He will have to give us comments on the studies so that we have the value of his input."

She agreed, arranging for Marty to fly up from Texas and review the data. I did not want copies of the data made and circulated, so it was necessary for Marty to come to my office and do his review right there. I fully expected that Marty would provide a report to CTIA on the data—which I believed would be a good thing. I knew this was going to be interesting, to put it mildly.

On February 1, 1999, Marty Meltz spent a full day in our conference room poring through the data—protocols, interim reports, and the most recent reports. He made it clear to me that he was doing this review as a consultant to CTIA. I made it clear to him that I saw this as part of the peer-review process, and that what he told CTIA about it was his business. He did his work. I did not see him before he left at the end of the day.

However, I did catch up with him the next day—much to my surprise. When I walked into the CTIA conference room to brief Tom and his top officials on the new data, Marty Meltz was there, sitting in a chair at the conference table. Wheeler was at the head of the table, of course. So I took a seat between the two of them, figuring I'd be closer to Wheeler—at least geographically.

CHAPTER ELEVEN

FOLLOW-THE-SCIENCE: *MORE RED FLAGS*

NO ONE—NOT Wireless Research Technology (WTR), not the government, and not the wireless phone industry—expected to see evidence of brain cancer caused by cellular phone use in the WTR studies that surveyed people's health problems. These were the epidemological studies, traditionally painstaking and time-consuming—and it was believed that cellular phones had not been in use long enough for any health risk as severe as tumors to show up in people. Further, given the uncertainties about what actually causes brain cancer and the imprecision of all epidemiological, or human, studies, the experts believed that only a very profound and widespread effect would be identifiable within the short period in which cellular phones had been in use.

Accordingly, the goals of the WTR human studies program were modest. Carlo simply wanted to lay a foundation for future tracking and monitoring of people who used wireless phones.

We saw our studies as establishing a baseline against which future data could be judged. In fact, Dr. Elizabeth Jacobson of the FDA told the CTIA's Board of Directors in January of 1997 that the tracking and monitoring—called post-market surveillance—should be considered by the wireless industry to be an ongoing cost of doing business. We saw our job as preparing the industry for the long haul, consistent with what the FDA had said they wanted.

But the findings of these first modest studies indicated a correlation between tumors and cell phones that had not been expected.

The WTR's Studies of People Using Cellular Phones

The holiday season of 1998 brought results from several studies by scientists working under contract to the WTR. One showed a statistically significant increase in cell phone users' risk of rare brain tumors at the brain's outer edge, on the side of the skull where cell phone antennas are held during calls.

The American Health Foundation (AHF) had contracted to do two studies for the WTR. One study was designed to look at the risk of brain cancer in cellular phone users—the question that had been on everyone's mind since 1993. The other study was designed to look at the risk of developing acoustic neuroma, which is a benign tumor of the nerve that controls hearing and extends from the ear into the brain. A WTR working group of cancer experts had identified acoustic neuroma as an important tumor for us to study because it is well within the physical range of the radiation plume that emanates from the antenna of a cellular phone.

Dr. Ernst Wynder and Dr. Gary Williams were well-respected scientists at the AHF. Carlo had previously worked with both physicians. Based on their involvement and the quality of the work they proposed, Carlo agreed to give AHF a contract to conduct these two studies. The two researchers began the work with a small grant from the U.S. Public Health Service, and the WTR's support would allow the study to be completed. Unfortunately, before the work was completed Dr. Wynder died and Dr. Williams left the AHF. The work was ultimately completed by Joshua Muscat, another AHF researcher, who was working on his doctoral degree.

A STUDY OF BRAIN-CANCER PATIENTS

Objective:

The objective of this study was to test whether the radiation from a handheld cellular telephone affects the development of brain cancer.

Procedure:

The work completed was a case-control study—patients with and without tumors were compared on the basis of whether, and how often, they used cellular phones. Their recalled responses were then compared to the responses of other patients in a control group. The patients were matched according to age, race, and gender so those variables would not affect the outcome of the analysis. The investigators gathered data for 466 newly diagnosed cases of primary brain cancer and 422 control subjects in five hospitals in the northeastern United States: Memorial Sloan-Kettering, New York University Medical Center, and Columbia University Presbyterian Hospital, all in New York City; Rhode Island Hospital in Providence; and Massachusetts General Hospital in Boston.

The study was two-thirds completed when we discovered it had a serious flaw: Most of the tumors studied were apparently outside the range within the skull in which radiation from a cell phone antenna would penetrate the brain. A separate effort by the WTR and others showed that radiation from a mobile phone antenna penetrates only a couple of inches into the head when the instrument is held against the ear. Thus, a large portion of the tumors in Muscat's study could not logically be linked to the cellular phone use because they simply were not relevant to a study of cell phone radiation. A more precise analysis based on the location of the tumors needed to be done. Thus, Muscat's planned overall analysis, which included all tumors regardless of their location in the brain, would be of little value in our effort to determine the impact of the cellular phone on brain cancer. But we conferred with Muscat and agreed that it might be beneficial if he could separate out a subgroup of tumors—those located at the side of the skull within penetration range of cell phone antenna emissions. He did this by studying the pathology report for each tumor and isolating tumors that were categorized as neuro-epithelial, tumors which are usually located at the sides of skulls. He then compared those cases in this subgroup with their control group (individuals who had no tumors).

Findings:

Not surprisingly, in the large overall group (which included mainly tumors outside the range of penetration by cell phone radiation), there was no increased risk of brain cancer among people who used cellular phones. The analysis of the control group showed that the procedure had indeed been valid; 18 percent of the control subjects reported using cellular phones, and that figure is very close to the estimated 20 percent of adults in the United States who had used cellular phones, according to surveys at the time.

However, a surprising finding did become evident when, for the WTR, Muscat analyzed that subgroup of neuro-epithelial tumors at the sides of individuals' heads. Of the 35 cases of neuro-epithelial tumors that he found, according to the final report that he submitted to the WTR, 14 were cellular phone users. Muscat compared that figure to the cell phone use among those in the control group (that had no tumors) and concluded that cell phone users were 2.4 times more likely to develop neuro-epithelial tumors. And that is indeed a statistically significant figure.

Significance:

This was the first cancer study that had ever been done on people who are cell phone users and it produced two significant findings: (1) In areas of the skull where radiation plumes from cell phones do penetrate, a significant increase in the risk of tumors appears evident. (2) for areas of the brain where cell phone radiation is now known not *to penetrate, cell phone usage does not seem to affect the risk of tumor development.*

While the positive finding in the end was based on just 35 cases, it is now imperative that a large study of all types of tumors near the sides of the head must be done promptly—for if the Muscat study is indeed a valid guide, a large study of this type of tumor can provide the most definitive evidence yet of whether cell phone radiation indeed causes brain cancer. A large international study of this magnitude was underway as of the year 2000, but it is not expected to be completed until 2005—a perilously long time for cell phone users to have to wait for information that can be so crucial to their health.

A STUDY OF ACOUSTIC NEUROMA PATIENTS

Objective:

The objective of this study was to assess the impact of cellular telephone use on the development of acoustic neuroma, a rare noncancerous tumor affecting the nerve that controls hearing.

Procedure:

This was also a case-control study in which 90 patients with acoustic neuroma were asked how often they had used their cellular phones over the previous years. Their responses were then compared to the responses of 86 patients of similar age, race, and gender who did not have tumors. The investigators gathered information from patients in three New York City hospitals: Memorial Sloan-Kettering, New York University Medical Center, and Columbia University Presbyterian Hospital.

Findings:

Of the 90 patients, 11 had used a cellular phone for three years or more. In the control group of 86 who did not have these tumors, only six had used a cell phone for three years or more. Therefore, those using cell phones for three years or more had a 60 percent greater risk of developing acoustic neuroma. The more years people used their cell phones, the greater was their risk of developing these benign tumors. Muscat's analysis concluded that the increased risk was statistically significant after patients had used cell phones for six years.

Significance:

Even though this study consisted of only a small number of tumors, it was enough to show a statistically significant dose–response relationship: The more years of cellular phone use, the higher the risk of developing acoustic neuroma.

> *The finding raises important questions about health risk from cellular phones and clearly points to areas that need more research. First, the acoustic nerve is within the 2-to-3-inch penetration zone of the radiation from the wireless phone antenna, so the nerve tissue definitely received exposure to radiation when patients used their cellular phones.*
>
> *There could be at least a couple of interpretations of the tumor-development importance that this study is really indicating. For example, the increased risk of acoustic neuroma for patients who used cell phones for six years could indicate the cumulative effect of radiation—that is, a person may need to use the phone for that many years to accumulate enough radio frequency radiation exposure for there to be a major problem. Or, the development of these tumors after six years of use could be an indicator of tumor latency—the effects of exposure to radiation may not begin to show up until six years have passed. Either of these possibilities is equally likely and each is supported by the data in the study.*

In addition, patients who entered the study in 1993 and 1994 and who had used cell phones for six years would have been using phones that were on the market during the late 1980s. At that time there were no guidelines with respect to radiation from cellular phones because these phones were exempted from FCC regulation. Radiation from these older phones was often much higher than exposures from some phones currently available. There was no regulation requirement, and no commercial incentive, for cellular phone manufacturers or service providers to keep the radiation levels low. Also, during the late 1980s and early 1990s, there were fewer base stations (the towers that relay the calls to and from individual phones). These phones had to work harder—use more power and thus emit more radiation—to carry a call from the base station to the phone that was being held against a user's head. People with a history of using these phones in those early years represent a high-exposure sentinel group that is very important for further study.

A STUDY OF 285,000 ANALOG CELLULAR PHONE USERS

On December 16, 1998, Dr. Ken Rothman and Dr. Nancy Dreyer, of Epidemiology Resources, Inc., of Newton, Mass., faxed the WTR Washington office a summary of their analysis of an epidemiology study they did for the WTR in which they examined 462 deaths that occurred in 1994 among a group of 285,561 cellular analog phone users.

This analysis was designed to be just the initial portion of a huge and vital ongoing study that would monitor almost a million cellular phone users. This first study was intended to provide analysis for a baseline year (1994); the large overall study was planned to cover a period of ten to 20 years. That would be a time frame long enough to span the known latent development period for cancers that could be related to cellular phone radiation—brain tumors, leukemia, tumors of glands in the neck—as well as noncancerous tumors of the head and auditory nerve.

The initial analysis by Drs. Rothman and Dreyer found six deaths attributed to brain cancer in that group of 462 people who died in 1994.

THE INITIAL ANALYSIS

Objective:

> The aim of this initial analysis was to see if there was an increased risk of dying from brain cancer among people who used handheld cellular phones (where the antenna is next to their heads) by comparing them to people who used car cellular phones (where the antenna is outside the car).

Procedure:

The type of research that Drs. Rothman and Dreyer and their colleagues completed is called a cohort mortality study. These studies are routinely used to assess dangers in the chemical, steel, and auto industries, among others. The concept was straightforward: to see if the pattern of causes of death among handheld phone users is different from the comparison group. For the study to be valid, it needed to account for all causes of death.

For this overall study to be successful, the investigating team had to be able to identify individuals who were indeed cell phone users. Then they had to be able to identify the type of cell phone used. So with the help of the WTR, the researchers recruited several of the largest cellular service provider companies, including Southwestern Bell and Cellular One, to participate in this study of cellular phone users by making available the companies' billing records for 770,390 customers in Boston, Chicago, Dallas, and Washington, D.C.

In all health-effect studies it is important to be able to start with basic demographic information of the individual cases. Thus, the companies had provided the research team with general demographic information on each customer—name, address, and phone number—that enabled the researchers to then discover which cell phone users had died in that year. The researchers worked with a credit bureau to get information on social security number, date of birth, and gender for those names gathered from the companies. The companies also provided the customer's phone serial number so researchers could determine the brand and type of phone used—whether it was a handheld or car phone. The companies provided information about the customers' billing history to determine how much the phone was used.

The initial analysis of 285,561 cell phone users came from an original data bank of 770,390 individuals. (People with duplicate Social Security numbers were eliminated, as were people whose Social Security numbers could not be verified. That left 316,084 records. An additional 75 records were eliminated because the account names suggested they were corporate accounts that many people might have used. The remaining names were matched against the Social Security Administration's Death Master File—a listing of all the people who died in the United States during 1994. Sixty-five percent of the people in the group studied were men. The median age of the people in the study was 42 years.)

Based on the makeup of the group of cellular phone users, rates of deaths from specific diseases and other causes were calculated based on national death statistics. The rates of death in cellular phone users were compared to rates in people who used only a car phone—with the antennas far away from their heads. The increase of health risk was then estimated by calculating the ratio of the death rates of the two groups.

Findings:

The study found that the rate of brain-cancer deaths was higher among handheld phone users, where the antenna was next to their heads, than among car phone users, where the antenna is far away from the user. The study found a total of six deaths from brain cancer in both groups. The study also found that, back in 1994, there were far more people who used car phones than used handheld phones; thus, four of the deaths occurred in the car phone group

> and two in the handheld group. By comparing those numbers with the type of phone used, the researchers calculated that the rate of death from brain cancer in users of non-handheld phones (car phones) was 2.42 per 100,000 people. The rate of death from brain cancer for handheld phone users, however, was higher: the calculated death rate for people who used handheld phones for three years or more at 8.42 per 100,000, a nearly fourfold increase.

Significance:

> *The overall study was designed to cover ten to 20 years. With just one year of data, it is impossible to determine whether the numbers in this analysis indicate the start of a pattern of increased risk of dying because of cellular phone use—a pattern that would become clear after subsequent years are analyzed—or whether this analysis for 1994 was just a chance occurrence. We needed death data for more years to put this one year finding into context.*

A ROADBLOCK

The WTR researchers needed more data, but were not able to obtain it.

After the Busse lawsuit was filed in the Chicago area, naming the WTR, Dr. Ken Rothman's group, and the cell phone manufacturers and carriers as defendants, officials at the Social Security Administration's National Death Index ended their cooperation with the research project. The director of The National Death Index decided he would not allow researchers access to death information for the years 1995 through 1998 until the legal proceedings were resolved. The plaintiffs in the lawsuit contended that the WTR study constituted human testing and an invasion of privacy, and said the researchers should have obtained informed consent from all of the people in the study before using the data. The researchers countered that obtaining informed consent from 700,000 people would have been nearly impossible logistically, and prohibitively expensive.

Meanwhile, the 1996 Telecommunications Act made it illegal for wireless phone carriers to release billing and other information about their individual customers—a provision that made it impossible for the researchers to continue their efforts. But Carlo and the WTR pressed their case before the FCC, emphasizing the research's vital importance to the health and safety of millions of cell phone customers. And so, after one year of appeals, the FCC granted the WTR a special research waiver.

But even in the fall of 2000, the Rothman-Dreyer research effort still was not able to proceed, because the National Death Index officials were awaiting the resolution of the Illinois lawsuit. And by that time the industry had ended its funding for the project.

CORROBORATION FROM SWEDEN

Eventually, in the year 2000, a Swedish epidemiological study provided new confirming evidence that the risk of tumors developing on the same side of the head that cell phone users hold their telephones is significantly higher than it is for the other side.

Dr. Lennart Hardell, of the Department of Oncology at Orebro Medical Center in Orebro, Sweden, directed a study that was requested by the World Health Organization as part of a multicountry epidemiological study of cancer and mobile telephone use. It was a case-control study designed to look at various risk factors for brain cancer—such as exposure to chemicals and to X-rays—and cellular telephone use was one of the factors included. The study asked patients with brain tumors, and a control group of people without brain tumors, about their phone use. Based on patients' responses to the questions, the investigators calculated cumulative use of cellular telephones, in hours, for each patient.

The hypothesis being tested was that use of the phone would increase the risk of brain cancer only in areas of the brain where there would be exposure to radiation from the cellular telephone antenna. The study therefore included only patients whose tumors were of a known anatomical location. Of the 217 patients with brain tumors, only 198 had anatomical location information—and of those, 136 were malignant tumors, 62 benign. All 198 of these cases were compared to their corresponding control patients who did not have tumors. A total of 99 patients had tumors on the right side of the brain and 78 had tumors on the left side; 21 patients had tumors in the middle of the brain.

The hypothesis was supported by the data, which showed that the risk of developing a brain tumor in the part of the brain that is on the same side of the head that a person holds a cell phone is significantly higher than the risk that the individual might develop a tumor anywhere else in the brain. Through a complex set of statistical calculations, the study concluded that, for cell phone users, the risk of developing a tumor in the area near the cell phone antenna was 2.4 times greater than the chance that a tumor would develop in any other portion of the brain. When other risk factors for brain tumors were controlled in a separate analysis, the risk of developing a tumor in the vicinity that a person holds a cell phone antenna was actually increased. The finding was statistically significant.

The Hardell study was consistent with the finding of the AHF, in an analysis done under a contract with the WTR. That AHF study found that the proximity of a cell phone antenna increases the risk of developing a tumor in that area of the head. Importantly, these two epidemiological studies employed very different methods, yet reached similar conclusions about the risk of cell phone users developing tumors.

VERY POLITICAL SCIENCE—III

At the insistence of the CTIA's Jo-Anne Basile, we agreed to have Dr. Linda Erdreich, a paid consultant of the CTIA, participate in the peer-review process of the three epidemiology studies. Dr. Erdreich's involvement was much the same as the involvement, at the CTIA's insistence, of Dr. Martin Meltz in the peer review of the genetic damage studies of

Drs. Tice and Hook. Becky Steffens-Jenrow, the WTR's epidemiology coordinator took the lead from our side. The deal was the same as with Marty Meltz—Dr. Erdreich could review the studies as part of the peer-review process, give us comments, and then she could report to the industry whatever she felt appropriate.

The first meeting took place on February 18th, 1999, in Dr. Erdreich's office at Bailey Research Associates in New York City. Becky and Erdreich met together for more than three hours, with Becky answering Erdreich's questions about the WTR program, the epidemiology component of the program, and the three completed studies by the Rothman group and the American Health Foundation.

Erdreich took copious notes, and then requested copies of the study reports. Becky called me from Dr. Erdreich's office to ask if that was OK. I said no, because the process was confidential, and until the peer-review board was finished with their review, I did not want the reports circulating. Becky left Erdreich alone with the reports for two additional hours, then came back to collect them. She gave Becky handwritten notes that were to go to the investigators as part of the peer review.

Becky transcribed the notes, prepared them for transmission to the investigators, and sent them to Erdreich for approval.

That meeting began a series of more than 15 personal telephone contacts between Becky and either Linda Erdreich or Jo-Anne Basile regarding specific points in the epidemiology studies. It seemed clear to me that a detailed approach to understanding these data was underway by the industry.

A second meeting took place on April 16th at Erdreich's office in New York. At this meeting, Erdreich again requested copies of the studies. By this time, the peer-review process was completed—and so, we obliged.

All told, Dr. Erdreich, as a paid CTIA consultant, reviewed and participated in the peer-review process for the WTR's epidemiology work for more than two months. She also had copies of the study and had repeated discussions with WTR investigator, Joshua Muscat. All of which made it very hard for me to understand why, the following autumn, Tom Wheeler would apparently try to discredit our work by claiming that we had never provided his CTIA with copies of these red-flag studies.

CHAPTER TWELVE

HOISTING THE RED FLAGS

AT ABOUT TEN MINUTES before 11:00, on the morning of February 2, 1999, George Carlo and Jeff Nesbit left the WTR townhouse office and walked one block west on N Street. It was a walk Carlo had made several times a week during the early years when he was a certified CTIA insider, only occasionally during the contentious middle years, and almost never during the last years of his roller-coaster relationship with Tom Wheeler. Carlo and Nesbit crossed Connecticut Avenue, entered the sumptuous office building at 1250 Connecticut, and walked to the main-floor conference room.

The conference room door was closed when we arrived. We knocked, and when the door opened, we walked in to silence. Seated at the large wooden conference table were Tom Wheeler; Margaret Tutwiler, CTIA's public relations vice president, who had served as Secretary of State James Baker's assistant secretary of state for public affairs during the George Bush administration; Jo-Anne Basile, CTIA vice president for health and safety, who had been in charge of riding herd on the WTR and me; and Brian Fontes, another CTIA vice president. They had apparently been talking with Dr. Marty Meltz, who, much to my surprise, was sitting near the middle of the table on the side nearest to the door. But they stopped talking when we came in. In fact, the only sound that had not ceased when we entered was from an overhead projector in the center of the table, facing the wall farthest from the door—it was turned on, but there was nothing being projected when we came in.

I suspected that they had been given a briefing by Meltz on the micronucleus studies before we arrived.

The pleasantries were short, and we got right down to business.

I passed out copies of the slides I was using.

I walked to the overhead projector and began with the results of the Rothman mortality study. The first slide detailed the results of their analysis of fatal automobile accidents among cellular phone users. I began:

"There is a statistically significant increase in the risk of deaths due to motor vehicle accidents among users of analog phones, regardless of whether the phone was handheld or with the antenna on the back window."

I was interrupted by Jo-Anne Basile. "Why did you include this analysis? This was not part of our Memorandum of Understanding. This is not your job."

"Jo-Anne, this is a cause-of-death study, and we have looked at all causes of death. This is the generally accepted approach for this type of epidemiology study. To leave it out would be improper. Besides, the auto-accident risk jumped out of the analysis as the most significant finding."

She replied. "We have our own approach to managing this issue. Your help is not needed."

As she spoke to me in front of Wheeler and Tutwiler, I realized this did not make her look good. She was supposed to be managing us, and this driving-safety matter had come as a surprise to her. Indeed, the CTIA had moved the WTR out of the driving-safety area of research by insisting upon the restrictive provisions of our Memorandum of Understanding. But it was there in the Rothman findings—and my job was to put it in front of Wheeler and the CTIA.

I continued. "There appears to be a dose–response for usage defined both by minutes of use per day and by number of calls per day: the more minutes and the more calls, the higher the risk."

I went on to explain that because this was the largest study done to date, with several hundred thousand person-years analyzed, it needed to be taken seriously. The data were quite persuasive.

When we looked at minutes of cellular phone use per day we saw the following: people who used their phones less than one minute per day had a fatal accident rate of about five per 100,000. When usage went up to three minutes per day, the rate went up to more than ten per 100,000. With usage greater than three minutes per day, the rate was 12.5. There were more than 300,000 person years analyzed in this part of the study—a very large number.

The calls-per-day analysis was a validation of the minutes-per-day study. Whenever two different measures suggest the same thing, it is more likely that what you are observing is real.

People who averaged less than a call every other day had a rate of death of about five per 100,000. When the usage went up to 1.5 calls per day, the rate went to almost 11 per 100,000—a statistically significant increase. At more than 1.5 calls per day, the rate went up to a statistically significant 13.4 per 100,000.

"The years-of-service analysis suggests that maybe over time, people get used to driving and talking on the phone. People who have been using their phones for less than three years have a significant increase in the risk of fatal accidents—about eight per 100,000—when compared to people who don't use phones. But after three years, the rate of death is no longer significantly high—down to about four per 100,000."

Jo-Anne jumped in. "There is no way to know if these people who died were using their cellular phone when they got into the accident, is there?"

"That is correct," I agreed. "The study was based on death certificates and that information is not included on a death certificate. But the findings are significant and this is a big study. They need to be looked at with the other studies that have been done on accidents."

She continued by suggesting that the minutes-per-day and calls-per-day studies were inconsistent with the years-of-service analysis. I disagreed.

"Dr. Rothman's suggestion seems to make the most sense: that over time, people become more accustomed to driving and talking, and that is reflected in the data."

I then presented the brain-cancer mortality data. "The rate of brain-cancer death is increased in handheld phone users. But it is based on only six cases—only on the deaths in 1994. We can't tell if this is a trend or an artifact, and won't be able to until we analyze what happened in '95, '96, and '97 with deaths in these people."

Tom Wheeler asked why this was not done in the first place. I reminded him of the Chicago lawsuit, and the reluctance of the National Death Index director to collaborate with us.

I then presented four slides summarizing what we had found in the case-control studies of acoustic neuroma and brain cancer.

By this time, Basile, Tutwiler, and Fontes were hurriedly taking notes.

Finally, I presented the genetic toxicology findings—with the very significant development of micronuclei when human blood cells were exposed to microwave radiation.

"There is evidence of genetic damage for all technologies after the 24-hour exposure," I said. "We are still trying to figure out what this means in light of the other studies showing no DNA damage."

Peculiarly, there was very little discussion about the genetic damage. I sensed that those discussions would take place with Dr. Meltz and without me.

When I was done, there was silence for what seemed like minutes. Wheeler sat back in his chair at the head of the conference table. He looked up at the ceiling with a pensive, exasperated look on his face. I could tell he was not happy about this.

He looked at me and said, "Looks like you need to pack for New Orleans. I want you to brief the CTIA board directly, next week." He did not want to carry this bad news to his board—I was to be the messenger. I thought back to the day in Wheeler's old, smaller office, in a different conference room, six years before. The project was little more than an idea, yet to be implemented. Tom had pointed at me and gave me a very clear message: ". . . I'm not going to be the fall guy if this goes bad—you are!"

He instructed his staff to return the copies of the overheads I had used. Jeff and I collected them and left the conference room.

TOUGH DAY IN THE BIG EASY

Wireless '99, the CTIA's trade show, was in full swing in New Orleans. The wireless industry's annual trade show has undergone a joyful explosion in size and influence under Tom Wheeler's leadership that has matched the incredible growth of the industry itself. As trade shows go, this 1999 gathering would have to be among the grandest. Entertainment by Tony Bennett and Wynonna Judd—and this year, a special convention-hall conversation with former U.S. President George Bush.

George Carlo arrived on February 9 for what he knew would be a very hard 24 hours in the Big Easy—a day of professional confrontation and, he feared, some not-so-professional whispering about him. He had spent the night tossing and turning, not really sleeping, and arrived at the convention hall at about 10:15 A.M. His presentation, delivering the troubling news to a private meeting of the CTIA board of directors—presidents, CEOs, and chairmen of the industry's top companies—was set for 12:30 P.M. With a couple of hours to kill, he walked into the auditorium to see what was on the program.

As I walked into the back of the large auditorium, I heard a familiar voice. Tom Wheeler was asking somebody a question. Then I heard another familiar voice: "Well Tom . . ." Former President George Bush was beginning his answer to his interviewer du jour, Tom Wheeler.

Far away, down in front of the huge, cavernous convention hall, I could see Bush and Wheeler on the stage, sitting in two chairs facing each other at 45-degree angles. Video cameras were capturing every facial expression and gesture. Bright lights made the stage look more like a movie sound stage than a business convention. Live closed-circuit feeds of the audio and video of this event were being shown in every corner of the convention center, for the benefit of those who wanted to be there in person but were too far away to see what was going on—unless they turned to their nearest TV set.

I thought to myself, Wheeler really knows how to do it up big. Everyone in the place was going to see Tom Wheeler talking to a man who had been the most powerful human being in the world. And of course, I was in New Orleans that day to bring the bad news to Wheeler's board.

Carlo did not stay to listen He ventured into the exhibit hall to see what was coming next in wireless technology. For an hour or so he just browsed, and thought through some of the points he would make to the CTIA board at 12:30.

It was now about 11:30, and at the AT&T exhibit, Bush and his enormous entourage were being treated to a briefing on the incredible "possibilities that PCS [personal communication system] phones offer for the new millennium." As they left the exhibit, Bush shook a score of hands, including mine.

At exactly 12:30 P.M., Carlo walked into the conference room where the board meeting was already underway. At a long table on the right side of the room, placed perpendicular to the podium, were several members of the CTIA staff. Mike Altschul, the CTIA's general counsel, rose to meet Carlo and led him to a seat against the wall, behind the CTIA staff table.

Jo-Anne Basile walked over to me and whispered that time was tight. "Cover what you need to in about ten minutes." She said the board had already been briefed and that they simply needed for me to hit the high points. The original plan had been for me to speak for about 30 minutes and then answer questions. To squeeze 30 minutes of data into ten minutes—and with not much time to think about it—was going to be a challenge. It caught me off guard, and I began to shuffle through my notes to re-do the talk. No time. Within ten minutes, Tom Wheeler was introducing me.

As I walked to the podium, I got the sense that the audience had indeed been briefed. I stood before them—about 70 people, I would guess now—and gave them the news, straight and to the point. It was the same information I had given Wheeler and his staff the week before. Except this had to be done in ten minutes—which meant that there wasn't any time for me to talk about our method and findings. I told them about our serious findings of genetic damage—the development of micronuclei, a finding of a biological effect that we had not expected, and which we had replicated several times to be sure there was no mistake.

As I ended, I reiterated, "We do not know what this means totally. The science is in a gray area, but there is more work that needs to be done. I have completed my commitment to you, but I will assist you as you move forward in whatever way I can."

Then I answered a few questions. Some centered on why I didn't tell them this in December; the answer was obvious: we did not have these data then, and I had made that very clear in my shortened presentation. It became equally obvious to me that a different story about me and my motives had already begun to circulate. And I would soon hear from friends and trusted associates that the word being spread was that I was simply looking for more research money. But, I had made it very clear to the board that I was not.

I then witnessed a discussion, led by Wheeler, about the need for the industry to do the right thing to sort through this science. A voice vote was taken. I later learned that what they voted on was never written down. But the sense of it was that the industry had voted unanimously to continue to do what was necessary—that is, provide the necessary funds—to complete the follow-on work that would confirm or refute these vital new findings of genetic damage, and to continue with more epidemiology.

A RARE EXECUTIVE: PROTECTING CUSTOMERS

William Collins, chairman of MetroCall, whose 120 stores across the country sell pagers and wireless phones, is not a member of the CTIA and was not in that board meeting—but he heard all about what Carlo had to say to the other industry chairmen and CEOs. "My brother-in-law was in that room, and he told me: 'We received some startling information from George Carlo.' He said George had laid out some significant concerns regarding cell phones and tumors and cancer. And that the most significant thing he remember from what George had said was: 'We've got to keep these phones out of the hands of kids.'"

The question of just how powerful Carlo's message had been that day turns out to be significant for just one reason: Tom Wheeler and the CTIA's website have taken the position that Carlo didn't have any urgent news to pass along that day and that he didn't speak to the board in a tone anywhere near as urgent as he used in public, on television, months later. MetroCall's Collins doesn't buy that. "From the way I heard about it, there was no mistaking George Carlo's message to the industry executives," Collins told Martin Schram.

Collins said he was indeed troubled by what he'd heard and later arranged to get a firsthand briefing from Carlo—and after that he put a new procedure into effect in his stores. Any customer who walks in is handed a single-page health and safety bulletin that explains the possible dangers of using cell phones. The sheet especially warns parents not to expose their children to health risks from cell phone radiation.

"Once we were provided with all of this information, we thought it was the right thing to do," said Collins, a rare breed of industry executive. "The one thing I did take away from everything that George pointed out was the effect it can have on kids. And we in the industry have to take the lead in getting that point across to the public—starting with our own customers. We've seen stonewalling by our industry association. We've seen hardball tactics by the CTIA—including attempts to discredit those who bring the warnings to the public. That's disheartening. "

CHAPTER THIRTEEN

When Science Collides with Politics

BACK IN WASHINGTON, George Carlo figured he would press on with his effort to follow the science to its logical public-health conclusion. He was not interested in—or really aware of—the politics. His responsibility, as he saw it, was to complete his investigation and make sure the findings were communicated to all who needed to hear them—the government regulators and Congress, the cell phone manufacturers and carriers, and the journalists who would carry the message to people everywhere who hold mobile phones next to their heads for minutes and even hours, every day of their lives.

But the more Carlo worked to assure that he had fully communicated the results of the science, the more rancorous his dealings became with the CTIA and its leader, the man who had hired him, sailed with him, and once even proposed buying a pleasure boat with him. Those days of smooth sailing on calm waters were over.

"New WTR Findings Raise Questions." The headline in the March 1, 1999, edition of *Radio Communications Report* (*RCR*), the respected wireless communications industry trade publication, was not the sort of thing that was greeted as good news in the halls of the CTIA. The report by *RCR*'s Washington correspondent Jeffrey Silva accurately reported that Carlo had informed federal regulators of five cell-culture studies raising questions of biological effects from mobile phone microwave emissions.

The news spread instantly through the cognoscenti of government and industry, and in time it made big news in the mass news media as well. Eventually, the *Washington Post* would print the news ("Study: Cell Phone Use May Have Cancer Link") and CNN's Steve Young would air a strong piece that included a sound bite from Carlo.

All the while, Carlo was discussing with Jo-Anne Basile the amount of money needed to follow up, as the CTIA board had overwhelmingly voted to do. Often he did his negotiating through Jeff Nesbit, who acted as a go-between. Two things were essential: Researchers needed to replicate, for a third time, the crucial micronucleus experiment to make sure that there could be no doubt about the findings that cell phone radiation can cause genetic damage in human blood; and they needed to convene an expert panel to evaluate the pathology in Joshua Muscat's epidemiological study. Nesbit reported to Carlo that the industry would give Wireless Technology Research (WTR) another $1.2 million to complete the follow-up work through the end of 1999. And that would then be the end of Carlo's involvement.

A MOST UNPLEASANT INTERLUDE

While George Carlo's business relationship with the CTIA was in turmoil, things were even worse for the public-health scientist on the home front. Carlo and his wife, Patricia, were getting a divorce—and it turned out to be a very contentious and bitter proceeding for both parties. Eventually, the Carlos' private discord would become a very public matter and cause considerable professional tumult. A couple of Christmas seasons earlier, Carlo had given his wife a most unusual present: stock that made her a half-owner of his public-health research company, Health Environmental Sciences Group (HES), where she would have the title of president. In the course of the divorce, her attorney filed court documents alleging that Carlo had committed fraud by taking funds from HES—without Patricia Carlo's knowledge—a leveraging tactic in the legal effort to secure the most lucrative settlement possible for her. Carlo, in turn, countered that such baseless allegations jeopardized his future lucrative contracts with the CTIA. This was his tactic aimed at securing the most favorable divorce terms for himself.

In March 1999 the CTIA was officially dragged into the Carlo divorce proceedings as Patricia Carlo's attorneys subpoenaed the CTIA to produce financial documents as part of a series of lawsuits, and to provide a CTIA official who would give testimony in a sworn deposition.

Late Thursday evening, Carlo's mobile phone rang while he was dining out. "George, this is Lisa [Joson, Carlo's assistant]. Tom Wheeler called and is demanding to see you tomorrow morning at 9:00 A.M."

"Did he say what he needed?"

"No, he didn't say. But he sounded serious."

"OK."

• • •

When I arrived at the CTIA office on Connecticut Avenue I was asked to wait for Tom in the reception area. There were two TV sets in the reception area. One was tuned to C-SPAN; the other was playing a video of the CTIA's promotional piece on safe driving: "Please use your phone safely while driving," was the main message. I had always wondered what they really meant by that. Did "use your phone safely" mean use a hands-free device so both hands could be on the wheel while driving—which made sense? Or were they just waffle words, meaning "just concentrate more"? If the message was to use a hands-free device, why not just say that? The whole campaign always struck me as more concerned with politics than exhibiting concern about safety.

Tom's secretary, Barbara Grant, came in and greeted me with a friendly smile as usual; but the air had completely changed from what it had been. In the beginning, in 1993, I was one of the gang, welcome just to walk right into Wheeler's inner sanctum, help myself to a cup of coffee, and shoot the breeze with his staff. Even a year ago, after the various controversies between us, I was still a welcomed visitor who was treated cordially, just a bit more formally. Now—quite understandably after all that had happened—I was just a guy waiting in the reception area. Barbara escorted me back to Wheeler's office and asked if I wanted coffee (I didn't), and I waited alone in the small conference room attached to Tom's office.

When Tom entered the room he had an expression of deep concern on his face.

"How are you doing, George?"

I was taken aback. His demeanor was more like a concerned friend than that of an adversary. But too much had gone on between us over the years for me to be open to a "just one of the boys" discussion. I was skeptical.

"Well, Tom, things have been better, but I'm making the best of it. Did you see the subpoenas?"

He nodded and said that was why he wanted to meet with me. He said the allegation about fraud had created a big problem for him. He said his board would crucify him if he gave me any more money, and that he thought this was pretty serious.

"There is no merit to those allegations, Tom, and you know it," I said. "The books have been audited continuously and you have seen where every penny has gone. Unfortunately, Patty and I are having a very nasty divorce."

"I understand," he said. "I've been there. I know how tough it gets." He said there was nothing he could do. He said he was going to do the follow-up research with someone else. He said maybe he would be able to "take care of" me in the fall with a consulting contract.

"Take care of" me. The message in that seemed to be that I'd get a consulting contract—if I didn't cause any more problems. He was using the subpoenas to gain an advantage. The board had approved

more scientific funding to complete the WTR's studies. But with the WTR and me tarred by these allegations, the $1.2 million could now be spent internally at CTIA. It seemed to me that we had been here before—in 1995, when the dollars earmarked for the WTR had been spent internally at CTIA.

I rose from my chair. "Goodbye, Tom."

By the end of 1999 the research that I heard the CTIA board vote to fund ten months earlier, based on my recommendations, still had not been done.

Later, Jo-Anne Basile was the person designated to testify in that deposition on behalf of the CTIA in answer to the subpoena obtained by Patricia Carlo's divorce attorney. In her sworn testimony, she was asked, "Are you aware if Mr. Wheeler had any conversations with Dr. Carlo concerning these allegations?" Basile replied, "I'm unaware of that." And indeed, she had not been in the room when Carlo and Wheeler had talked about that very topic.

Carlo's private problems would become public knowledge that summer. *RCR*'s Jeffrey Silva says he received a tip that the Carlo divorce had turned nasty and that charges of fraud and stealing money were contained in court filings. Silva, being a diligent reporter, did his job and dug out the court papers, read them for himself, and then wrote a story detailing the allegation. It was newsworthy for the trade publication because it concerned the man who headed the WTR, but the allegation actually concerned another organization altogether, HES, owned by Carlo and his wife. As happens in most bitter divorce proceedings, all of the nasty charges and countercharges were promptly dropped as part of the final divorce settlement. But the damage was done. Silva's accurate report of the unsupported allegation of fraud had a devastating impact on Carlo's public and professional image.

• • •

During the months of March and April in 1999, an extensive peer-review process was underway: About 30 different scientists, in academia and in government, received copies of the papers detailing the findings of Ray Tice and Graham Hook's test-tube experiments that had shown the development of micronuclei in human blood exposed to cell phone radiation, and Muscat's epidemiology studies suggesting a surprising increase in tumors by cell phone users. The scientists were asked to provide the WTR with critical review and comment.

Apparently, one of that group talked to John Schwartz of the Washington Post *about the findings. I received a call from Schwartz in early April; he indicated he had heard that we had some interesting findings and he wanted to do an interview. He said he had already talked to someone in the government. Because the peer-review process was not complete, I was not comfortable talking to him, though I told him I would be happy to meet with him when we had the peer review completed in the middle of May. We set an appointment for May 14th. But before I got off the phone, I asked him what he had heard. He had it all.*

In May we conducted the interview, and the article ran on Saturday morning, May 22. Schwartz called me the day before to alert me, but told me he was a little disappointed that the article had been shortened extensively and moved from Sunday to Saturday, when most people are running errands to the hardware store rather than reading the paper.

Saturday morning my phone rang before 8:00 A.M.

"George," the voice on the phone said angrily. "This is Tom. Have you seen the Post.*"*

"No, I haven't, but if you wait a minute, I'll get it from my front porch," I said.

"I'll wait."

I read the article while Wheeler steamed on the other end of the phone.

The story was in the business section, on page E-1, beneath the headline: "Study: Cell Phone Use May Have Cancer Link." The story had accurately reported: "The data, while 'important,' only suggest that more research is necessary, said George Carlo, chairman of the industry-funded Wireless Technology Research group. 'We're now in a gray area that we've never been in before with this. When we're in a gray area, the best thing to do is let the public know about the findings so that they can make their own judgment,' he said." At the end of the piece, it said: "Carlo, who uses a cellular phone with a plug-in earpiece that allows him to talk without holding the device to his head, said he chose to issue the results before scientific publication so that government and industry could take the next step in research. 'What we don't want to do two, three, four years from now is to say, 'God—this was the tip of the iceberg, and we didn't see it.'"

Wheeler was livid. "What are you going to do about this, George? That bit about the headset is a real gem. Very damaging. We are going into the office to issue a release right away to straighten this out. You should be doing the same."

I couldn't believe it. Wheeler was upset because people would know that I used a headset! Apparently giving the people information on how they can be sure they are safe was bad business. I played golf on Saturday. Monday morning I wrote and issued a statement.

• • •

One of the sharpest and most complete reports on Carlo's one-man intervention on behalf of cell phone consumers appeared in the *Boston Globe* on October 4, 1999. Reporter Patricia Wen's article began:

> Almost nobody expected George Carlo, of all people, to be warning consumers about the possible dangers of cell phones.
>
> Back in 1993, Carlo was dubbed "industry's boy" by consumer advocates. . . . But now that the project is winding down and its final report is due later this year, Carlo has created a stir by saying that consumers should take some precautions when using cellular phones, even while scientists at the US Food

> and Drug Administration and elsewhere say that cell phones do not pose any danger to users.

As a sidebar to this main story, the *Globe* responsibly ran a lengthy question-and-answer transcript taken from Carlo's interview with Wen, under the headline: "A Controversial Call on Cell Phone Use." In the interview, Carlo actually maintained a relatively cautious public posture. For example:

> Q: Do you believe today's cell phones pose some danger to users?
> A: The science is in a gray area. We have scientific information now that suggests genetic damage and some increased risk of cancer. My current recommendation at this point has to do with moving the antenna away from your head.

Later that year, ABC News' Brenda Breslauer, a producer for the network's *20/20* news magazine program, began reporting for a major piece that would air in the fall, with Brian Ross as the correspondent. Word of a prospective prime-time report about cell phone health risks sent Tom Wheeler into his commander-in-chief mode, as he launched another of his battle-tested preemptive strikes.

On October 11, 1999, there arrived in the Manhattan office of David Westin, president of ABC News, a letter from W. Andrew Copenhaver, a Washington lawyer representing the CTIA who was already well known at ABC—he'd represented the Food Lion supermarkets in their newsmaking lawsuit against the network. "I am writing on behalf of the Cellular Telecommunications Industry Association concerning your forthcoming *20/20* broadcast on wireless phones," the letter began. The three-page letter went on to list a number of the industry's concerns about what ABC News would be reporting—and take a slap at Carlo's motives. The letter noted that Carlo had recently published a book (actually a ring-bound notebook on the risks posed by cell phones; its consumer recommendations can be found in Chapter Eighteen of this book). The letter maintained to ABC that Carlo "is seeking personal advantage from his statements to *20/20*"—statements that, of course, ABC News had not yet aired and the CTIA had not yet heard. The final paragraph of the lawyer's letter contained the CTIA's actual goal: "I ask that you or your designees personally get involved in determining if *20/20*'s proposed program on wireless phones meets ABC's journalistic standards, and that you delay airing of the program until the review is complete."

Indeed, the CTIA wound up making more news for itself: "CTIA Attempts to Delay TV Show," said the October 18, 1999, headline in the trade publication *RCR*. On the night of October 20, 1999, ABC News' *20/20* aired its story—and in it, Carlo said his piece directly to cell phone users who were most in need of some straight talk and simply were not getting it from the industry or the regulators. The ABC piece gave Wheeler an early opportunity to make his case.

> **BRIAN ROSS** Thomas Wheeler is the president of the cell phone industry's trade group in Washington, D.C.
>
> **THOMAS WHEELER** Our industry has gone out and aggressively asked the question, "Can we find a problem?" And the answer that has come back is that there is nothing that has come up in the research that suggests that there is a linkage between the use of a wireless phone and health effects. . . .
>
> **DR. LOUIS SLESIN** Nonsense, in a word. Simple nonsense.
>
> **ROSS** Dr. Louis Slesin is the editor of *Microwave News*, a widely read and influential trade newsletter which tracks the cell phone business and frequently criticizes what Slesin says is the industry's attempt to ignore or spin troublesome scientific findings.
>
> **SLESIN** This is about PR, not about science. There's research from Australia, there's research from England, Denmark, Sweden, Norway, Germany, all pointing in a direction Mr. Wheeler

doesn't want to look. Essentially, we have reports of headaches, of cancer, of changes in blood pressure, changes in sleeping patterns. . . . .

ROSS [Ross noted that the national "scare" began with CNN's 1993 Larry King piece on a cell phone–cancer lawsuit.] And [that] led to the announcement of a $25-million industry research program to be run by Dr. George Carlo, a public-health consultant, who was labeled then by some as a kind of scientific shill for the cell phone industry. Do you think they thought they had bought you?

GEORGE CARLO I—I hope that they didn't, but I think that they probably did.

ROSS And, now, after six years of running the industry's research program, Dr. Carlo has come to a surprising conclusion, forcing him, he says, to break ranks with the industry to add his voice to those increasingly concerned about the safety of cell phones.

CARLO We've moved into an area where we now have some direct evidence of possible harm from cellular phones.

ROSS In a revealing interview with *20/20*, Dr. Carlo said he felt he had no choice but to blow the whistle on what he says has been going on behind the scenes.

CARLO The industry had come out right after that program and said that there were thousands of studies that proved that wireless phones are safe, and the fact was there were no studies that were directly relevant.

ROSS Meaning, no studies directly relevant to cell phone exposure. But there are now, including studies Carlo oversaw and that the industry approved and paid for . . . clearly suggesting two potential problems, according to Carlo. Genetic damage, based on laboratory tests involving human blood, and an increased risk of a rare type of brain tumor, based on a study of brain tumor patients, although no overall increase in cancer was found.

CARLO The type of tumor is consistent with the idea that it's—it could be affected by the radiation coming from the antenna.

ROSS But if these phones were so bad, wouldn't we be seeing thousands, tens of thousands, of people with brain tumors right now?

CARLO Not necessarily. The—the technology has not been around that long and cancer is a disease that has a long latency period. It usually takes ten to 15 years for tumors to develop.

ROSS . . . some of Dr. Carlo's scientific colleagues, including the author on the brain tumor study, disagree with Carlo's interpretation of the findings. One of them is Dr. Martin Meltz, a scientist at the University of Texas and a paid industry consultant whom the industry said we should talk to.

DR. MARTIN MELTZ I believe, from my perspective, that the weight of knowledge indicates safety of cell phone use.

ROSS But Carlo says the new studies, while not proving cell phones are dangerous, do contradict such assurances that cell phones are safe. And that's something the industry knows? You've shown them these same slides?

CARLO That's correct.

ROSS The cell phone industry also sought to down-play Dr. Carlo's stunning defection with this formal statement, saying, quote, "The prevailing scientific consensus is that there is no evidence of risk from the use of wireless phones." No evidence of risk. Is that true?

CARLO That's wrong.

ROSS That's wrong?

CARLO That's wrong.

ROSS Have you seen this?

CARLO It's actually quite shocking, knowing—knowing what has been conveyed to them . . .

ROSS Other scientists we checked with also took sharp exception to the industry's position about no evidence of risk . . . Even the scientist the industry told us to talk to, Dr. Meltz, reluctantly conceded there is some evidence that needs follow-up.

MELTZ There is evidence. I have to say that, now, I—I—there is evidence of risk. Whether it is valid evidence of risk or not needs to be further examined . . .

ROSS Aren't you concerned when you hear those possible health effects . . .

WHEELER I have . . .

ROSS . . . brain tumors, genetic damage?

WHEELER . . . I have to look at what the responsible scientists say . . .

ROSS They're alarmed by this.

WHEELER . . . and—and they say that there is not a public-health effect . . .

ROSS Who are you say—who says that?

WHEELER . . . and—and they say . . .

ROSS Who actually says that?

WHEELER This is—this is what they—what the FDA has said.

ROSS Not exactly. When we checked the website of the FDA, the Food and Drug Administration, we found a much more qualified position on cell phones. The FDA says, while the available science does not demonstrate harm from cell phones, [neither] does it lead to the conclusion that they are absolutely safe.

• • •

On October 20, 1999, Carlo's WTR group prepared a statement that wasn't really *news* at all. It merely noted that back in 1994 the WTR's predecessor group, the Science Advisory Group (SAG), had urged the adoption of three crucial recommendations—which the industry and the FDA had approved—but the regulators failed to enact them or even monitor what the industry was doing to assure that the public interest was being served. The recommendations all dealt with informing the public about possible health effects:

1. Adopt standardized labeling of wireless instruments;
2. Develop standardized information for dissemination to [wireless] companies and to the public; and
3. Adopt an industry-wide instrument certification

> program that requires certified phones to meet all appropriate standards.

The three recommendations were an essential beginning step laying the groundwork that would have made possible a future intervention in the event that a health risk was identified. The industry had sent the FDA a letter, dated December 9, 1994, concurring with the recommendations.

But nothing of substance had ever been done—not by the industry, and not by the regulators who could have seen to it that these three simple, common-sense recommendations were implemented. And there was one further shortcoming: The news media had failed to pursue the issue—failed to push the industry and its regulators to explain why there had been no action on this subject in the intervening five years.

A LETTER TO THE INDUSTRY CHIEFS

Now there could be no turning back, no need to search for a safe middle ground, no hope that tensions could be smoothed over if Carlo would only downplay the true significance of what he'd learned and tell the industry's leaders only what they most wanted to hear. Carlo was about to take his boldest step of all. In October 1999 Carlo wrote 28 identical letters on WTR stationery and sent them to the chairmen and CEOs of the cellular telephone industry. They were the ones who had paid for his six-year effort and they deserved to hear, directly from him, his candid assessment of the findings their studies had produced.

• • •

> 7 October 1999
> Mr. C. Michael Armstrong
> Chairman and Chief Executive Officer
> AT&T Corporation . . .
>
> Dear Mr. Armstrong:
>
> After much thought, I am writing this letter to you, personally, to ask your assistance in solving what I believe

is an emerging and serious problem concerning wireless phones. I write this letter in the interest of the more than 80 million wireless phone users in the United States and the more than 200 million worldwide. But I also write this letter in the interest of your industry, a critical part of our social and economic infrastructure.

Since 1993, I have headed the WTR surveillance and research program funded by the wireless industry. The goal of WTR has always been to identify and solve any problems concerning consumers' health that could arise from the use of these phones. This past February, at the annual convention of the CTIA, I met with the full board of that organization to brief them on some surprising findings from our work. I do not recall if you were there personally, but my understanding is that all segments of the industry were represented.

At that briefing, I explained that the well-conducted scientific studies that WTR was overseeing indicated that the question of wireless phone safety had become confused.

Specifically, I reported to you that:

- the rate of death from brain cancer among handheld phone users was higher than the rate of brain cancer death among those who used non-handheld phones that were away from their head;
- the risk of acoustic neuroma, a benign tumor of the auditory nerve that is well in range of the radiation coming from a phone's antenna, was 50 percent higher in people who reported using cell phones for six years or more; moreover, that relationship between the amount of cell phone use and this tumor appeared to follow a dose-response curve;
- the risk of rare neuro-epithelial tumors on the outside of the brain was more than doubled, a statistically significant risk increase, in cell phone users as compared to people who did not use cell phones;
- there appeared to be some correlation between brain tumors occurring on the right side of the head and use of the phone on the right side of the head;

- laboratory studies looking at the ability of radiation from a phone's antenna to cause functional genetic damage were definitively positive, and were following a dose-response relationship.

I also indicated that while our overall study of brain cancer occurrence did not show a correlation with cell phone use, the vast majority of the tumors that were studied were well out of range of the radiation that one would expect from a cell phone's antenna. Because of that distance, the finding of no effect was questionable. Such misclassification of radiation exposure would tend to dilute any real effect that may have been present. In addition, I reported to you that the genetic damage studies we conducted to look at the ability of radiation from the phones to break DNA were negative, but that the positive finding of functional DNA damage could be more important, perhaps indicating a problem that is not dependent on DNA breakage, and that these inconsistencies needed to be clarified. I reported that while none of these findings alone were evidence of a definitive health hazard from wireless phones, the pattern of potential health effects evidenced by different types of studies, from different laboratories, and by different investigators raised serious questions.

Following my presentation, I heard by voice vote of those present, a pledge to "do the right thing in following up these findings" and a commitment of the necessary funds.

When I took on the responsibility of doing this work for you, I pledged five years. I was asked to continue on through the end of a sixth year, and agreed. My tenure is now completed. My presentation to you and the CTIA board in February was not an effort to lengthen my tenure at WTR, nor to lengthen the tenure of WTR itself. I was simply doing my job of letting you know what we found and what needed to be done following from our findings. I made this expressly clear during my presentation to you and in many subsequent conversations with members of your industry and the media.

Today, I sit here extremely frustrated and concerned that appropriate steps have not been taken by the wireless industry to protect consumers during this time of uncertainty about safety. The steps I am referring to specifically followed from the WTR program and have been recommended repeatedly in public and private fora by me and other experts from around the world. As I prepare to move away from the wireless phone issue and into a different public health direction, I am concerned that the wireless industry is missing a valuable opportunity by dealing with these public health concerns through politics, creating illusions that more research over the next several years helps consumers today, and false claims that regulatory compliance means safety. The better choice by the wireless industry would be to implement measured steps aimed at true consumer protection.

Alarmingly, indications are that some segments of the industry have ignored the scientific findings suggesting potential health effects, have repeatedly and falsely claimed that wireless phones are safe for all consumers including children, and have created an illusion of responsible follow up by calling for and supporting more research. The most important measures of consumer protection are missing: complete and honest factual information to allow informed judgement by consumers about assumption of risk; the direct tracking and monitoring of what happens to consumers who use wireless phones; and, the monitoring of changes in the technology that could impact health.

I am especially concerned about what appear to be actions by a segment of the industry to conscript the FCC, the FDA and The World Health Organization with them in following a non-effectual course that will likely result in a regulatory and consumer backlash.

As an industry, you will have to deal with the fallout from all of your choices, good and bad, in the long term. But short term, I would like your help in effectuating an important public health intervention today.

The question of wireless phone safety is unclear. Therefore, from a public health perspective, it is critical for consumers to have the information they need to make an informed judgement about how much of this unknown risk they wish to assume in their use of wireless phones. Informing consumers openly and honestly about what is known and not-known about health risks is not liability laden—it is evidence that your industry is being responsible, and doing all it can to assure safe use of its products. The current popular backlash we are witnessing in the United States today against the tobacco industry is derived in large part from perceived dishonesty on the part of that industry in not being forthright about health effects. I urge you to help your industry not repeat that mistake.

As we close out the business of the WTR, I would like to openly ask for your help in distributing the summary findings we have compiled of our work. This last action is what always has been anticipated and forecast in the WTR's research agenda. I have asked another organization with which I am affiliated, The Health Risk Management Group (HRMG), to help us with this public health intervention step, and to put together a consumer information package for widespread distribution. Because neither WTR nor HRMG have the means to effectuate this intervention, I am asking you to help us do the right thing.

I would be happy to talk to you personally about this.

Sincerely yours,
George L. Carlo, Ph.D., J.D., M.S.
Chairman

• • •

The "consumer information package" that Carlo referred to in the last paragraph of his letter to the industry CEOs and board chairmen was, in fact, a bit more than just that. Enclosed in a blue plastic ring-bound notebook, the contents were intended as a consumer protection guide, which Carlo's new company, Health Risk

Management Group, was advertising on its website for sale to consumers at $19.95 each. It gave Wheeler an opening to attempt to discredit Carlo at every opportunity. Wheeler's assistants began telling journalists that it seemed as though Carlo was out to exploit his findings to make a buck—by making provocative public statements about health risks just to sell what they called his "book." A number of journalists later said privately that, after CTIA officials brought it to their attention, they began to wonder if Carlo's main motive was public health, private profit, or both.

Looking back, it was a mistake for us to have tried to sell that consumer information packet and market it on our website. It gave our industry critics an opportunity to cast aspersions on all of our motives, and it may even have made some journalists view skeptically findings that needed to be followed up, scientifically and journalistically. As I saw it, the latest scientific findings meant that this was an urgent problem—and the industry was not going to fund the research in time to benefit consumers. If we were going to do it ourselves, we needed the money that a consumer packet might have raised. But in retrospect, I think people shouldn't have to pay to learn this vital health information. Ideally, the government should have provided it to the public. The industry should have, too. Regardless, we just should have just put the information up on our website—for free.

Carlo's letter to the top brass of Wheeler's industry—Wheeler's bosses—apparently did not go down very well with the CTIA commander-in-chief. The day after Carlo's letter was faxed to the CEOs and board chairmen, Wheeler wrote a scathing three-page rejoinder to Carlo, who got it via the post office several days after the trade press journalists had received their copies, and had begun calling to ask Carlo about it. Wheeler opened by referring to the letter to the CEOs and board chairmen not as a plea for action but as a promotion for what he called Carlo's "new book." And after years of referring to Carlo in letters to the FDA, the SAG, and the WTR as "Dr. Carlo," Wheeler now chose to refer to his former handpicked research chief (who has a doctorate in pathology) as "Mr. Carlo."

Wheeler's letter, dated October 8, 1999, began by stating that the media had "shared with us" a copy of his press release and the "letter you sent out promoting your new book"—which was (actually a loose-leaf notebook) telling consumers how they can protect themselves.

The CTIA president went on to say that "we are certain that you have never provided CTIA with the studies you mention in your letter." He said he didn't think Carlo "withheld them on purpose, but believed they were not complete. "If you now have specific and complete scientific data, then we respectfully request that you immediately provide it in its entirety to us as well as the world's scientists for their review," Wheeler wrote.

Not only had Carlo not withheld anything from Wheeler and others in the industry, he made sure they had the latest and best information on the new and troubling findings. Back in February, Carlo had felt the new red-flag findings were so important that he notified Wheeler and his staff about them immediately. He gave Wheeler a briefing, complete with slides, on all of the preliminary findings. And then, throughout the peer-review process, Carlo had arranged for two scientists of Wheeler's choosing—biologist Martin Meltz and epidemiologist Linda Erdreich—to participate in the peer-review process with the authorized understanding that they would report back to Wheeler. Throughout the entire history of the WTR, studies were deemed final at the completion of the peer review. That is when the findings were made public, as was the case, for example, in the pacemaker interference study, when recommendations from the work went into effect eight months before the study was published in the *New England Journal of Medicine*.

Here is a detailed listing of how and when Wheeler and his designated representatives were kept informed about every detail of every finding:

1. Private briefing to CTIA scientific consultant Martin Meltz on February 1, 1999; he reviewed all data on genetic damage (in the form of micronuclei) and asked for permission and was granted it to give CTIA notes on the briefing.

2. Private briefing by Carlo to CTIA executives February 2, 1999, at Tom Wheeler's headquarters, complete with a slide presentation.

3. Briefing for the government Interagency Working Group at FDA headquarters in Rockville, Maryland, on February 9, 1999, attended by CTIA officials.
4. Briefing by Carlo to CTIA Board of Directors in New Orleans, February 10, 1999.

5. Epidemiology data extensively reviewed by CTIA consultant Linda Erdreich in April 1999; the complete final reports of the studies were given to Dr. Erdreich after peer review was completed in May 1999.

6. CTIA's consultant Martin Meltz participated in special peer-review meeting where genetic damage (micronuclei) findings were presented in May 1999.

7. Peer review was completed in May 1999; study results were deemed ready to be publicized at that time, according to the procedure established by the CTIA and WTR at the inception of the program in 1993.

8. CTIA participated in State of the Science Colloquium in Long Beach, June 19–20, 1999, where the scientists who conducted the studies presented the now–peer-reviewed findings.

9. CTIA published on its website the full text of the reports made by the scientists at the Colloquium in the summer of 1999, after having requested and received from Carlo the complete audiotape of the proceedings, from which the CTIA then made its detailed transcript.

10. CTIA acknowledged that it had received the abstract of the epidemiological studies of conducted by Muscat. In comments published in the Federal Register on March 2, 2000, on nominations for toxicological studies to be undertaken by the National Toxicology Program on substances of potential human health concern, the CTIA stated: "The Muscat study involved newly diagnosed cases of brain cancer from five U.S. hospitals and was designed to look at both duration and frequency of cellular telephone use. This study has not been published to date, but a copy of the abstract presented at the Colloquium is included with these comments."

Finally, in February 2000 the CTIA received the final WTR report that summarized all of the findings—all of which had been peer-reviewed the previous May, and all of which had been conveyed previously to the trade association, the industry, the government, and most importantly, the public. In the summer of 2000, Carlo published in the peer-reviewed online medical journal *Medscape*, an extensive review of all of the science, including detailed accounts of the WTR-contracted *in vitro* studies which found genetic damage and epidemiological studies that suggested an increase of cancer and other health risks from cell phones.

Yet, on the CTIA's website, and in interviews with journalists, the CTIA officials noted repeatedly, throughout 1999 and 2000, that the most alarming findings cited by Carlo still had not been formally published in recognized scientific journals.

There is indeed one valid reason why the findings from peer-reviewed studies, done by recognized scientists from around the country, were not published as soon as they were completed. It was because the CTIA had ended its funding of the program and there were no funds to pay these distinguished scientists for the additional time—perhaps several months—it would have required to prepare their mate-

rials for publication. The WTR contracts with the scientists ended with the completion of peer review.

CTIA used the fact that the studies hadn't been immediately published as a powerful public-relations weapon. With that weapon the CTIA lobbyists were able to create doubts among journalists from the major mass-media news organizations as to whether there might be flaws or other problems with the studies and their findings. Journalists lost sight of the fact that the CTIA had never waited for publication of findings in the earlier years, when its lobbyists rushed to spread word of findings that found no health problems.

My biggest regret is that I didn't realize the PR war that was being waged until it was too late. I didn't make the push for quick publication a top priority—because the studies had already been peer-reviewed and we were using the results to help consumers protect their health. Meanwhile, the scientific process was continuing at the same pace it always has; publication frequently lags years behind findings. My stubborn reliance upon the completion of peer review as more important than publication backfired politically.

I wish now I had been more savvy. The smart PR ploy would have been to put the ball right in the CTIA's court by writing a one-paragraph letter telling Tom Wheeler that since the industry views it important to the public interest that these findings be published immediately, then all the industry has to do to serve this public interest is pay each of these scientists directly the small amount of money it would have required to get their findings ready for publication sooner. All of the studies could have been prepared for just about $100,000—a pittance for a multibillion-dollar industry that says it is only trying to inform the public. And for that matter, I could have written the FDA and suggested that somewhere in the bureaucracy surely there must be funds to pay for the publication. Either way, once those letters were made available to journalists they would have understood just what—and who—was holding up the publication process.

Instead, the industry found me an inviting target, casting aspersions upon my credibility through its website and in its interviews with journalists. They tried tarnishing me to raise doubts about the findings of these reputable scientists. It's clear to me now that from the day Wheeler and I first shook hands, we were on a collision course, but I didn't see it coming.

CHAPTER FOURTEEN

Protect the Children

CHILDREN AND PARENTS could not miss the lure of the full-page, full-color newspaper ads: There in glorious reds and pinks and blues and greens were the grinning faces of Mickey and Minnie Mouse and Goofy—each adorning a brand-new Nokia wireless phone.

And children and parents could not miss the pitch: Stretched as wide as the entire newspaper page was a three-inch band of red, emblazoned with white letters: "Admit it, you want one."

"Right now, give everyone on your list a little Disney holiday magic," the ad said. "Choose from four Disney Xpress-on color covers . . . Mickey, Minnie, Donald, or Goofy make the perfect stocking stuffers. Only from AT&T."

AT&T paid grandly to run these full-page ads in newspapers across the United States in the holiday season of 1999—on November 12 and 19 in the *New York Times*, for example. To further lure consumers into buying these bright-and-happy Disney cartoon phones for their children, the ads promised a $30 Nokia rebate to all who signed up for the AT&T Family Plan. It was a costly but carefully planned holiday-season ad blitz in which AT&T enlisted Disney characters and Nokia in their effort to push the AT&T "family plan" wireless service into American homes everywhere; and push these Nokia phones into young hands everywhere—young hands that would be holding these phones to their very young heads.

"Admit it, you want one"—AT&T's Disney Nokia ad campaign set a new benchmark for commercial cynicism. For what the manufacturers and carriers were not admitting in their bright-and-happy—and enticing—ads was that three years earlier, scientific information had circulated widely through the wireless telephone industry that constituted an unmistakable warning: The radiation plume that emanates from a cell phone antenna penetrates much deeper into the heads of children than adults. And, once, it penetrates children's skulls it enters their brains and eyes at an absorption rate far greater than it does in adults.

Dr. Om Gandhi, a highly respected scientist at the University of Utah, did that study in 1996 and found that the differences in the rates of penetration into the heads of five-year-old children, ten-year-old children, and adults were especially shocking. Gandhi's study compared the average specific absorption rate of radiation (measured in milliwatts per kilogram, or mW/kg) in the three age groups. Gandhi found that the radiation absorption rates inside the brain (in mW/kg) were:

* 7.84 in an adult
* 19.77 in a ten-year-old child
* 33.12 in a five-year-old child.

Radiation absorption rates in the fluid of the eye were:

* 3.3 in an adult
* 18.38 in a ten-year-old child
* 40.18 in a five-year old child

Radiation absorption rates in the lens of the eye were:

* 1.34 in an adult
* 6.93 in a ten-year-old child
* 15.6 in a five-year-old child

> Finally, radiation absorption rates in the connective tissue of the eye were:
> * 1.77 in an adult
> * 9.8 in a ten-year-old child
> *19.69 in a five-year-old child

These differences in exposure are profoundly large, and signify potentially serious health risks to children from radio waves—risks far more serious than for adults. This concern also must apply to genetic damage and cancer. The current science strongly suggests that genetic damage is associated with exposure to radio waves from the antennas of mobile telephones. Most alarming are the very consistent data worldwide showing micronucleus formation in blood cells following radio wave exposure.

A WORLD OF CONCERN

There is worldwide agreement that the question of micronucleus formation should be at the research forefront. The British Parliament's Independent Expert Group on Mobile Phones, also called the Stewart Commission, in a comprehensive summary of research addressing mobile phones and health, released in May 2000, singled out the studies of micronuclei formation after exposure to radio frequency radiation as a consistent finding that needs careful follow-up research. The Royal Society of Canada concluded the same in their report on peer-reviewed studies that was published in 1998. Most recently, the U.S. FDA convened a group of experts to look at the same problem and to recommend scientific follow-up research.

The concern is that children are more susceptible to genetic damage because the tissues in their brains and bodies are still growing and their cells are rapidly dividing. Damage to the genetic material in growing cells can lead to disruption of cellular function, cell death, the development of tumors, and damage to the immune and nervous systems. Further, the protective systems that allow for adaptation to environmental insults of all types are not fully developed in children. In the brain, for example, these systems develop and grow until a person is in his or her early 20s.

In itself, the higher susceptibility of children and teenagers to the types of health risks that the radio wave scientific data are now showing would be reason to add additional protections for children from the effects of radio waves.

But the situation is much more serious. This increase in susceptibility coupled with the significantly higher penetration of radio wave radiation into children's heads, brains, and eyes calls for immediate action to protect children.

The Stewart Commission recommended such steps. They concluded:

> "If there are currently unrecognised adverse health effects from the use of mobile phones, children may be more vulnerable because of their developing nervous system, the greater absorption of energy in the tissues of the head, and a longer lifetime of exposure. In line with our precautionary approach, we believe that the widespread use of mobile phones by children for non-essential calls should be discouraged. We also recommend that the mobile phone industry should refrain from promoting the use of mobile phones by children."

These concerns also reach to the new and expanding wireless Internet, which will use the same radio-wave transmission technology as mobile telephones. Among the knowledgeable experts who have raised concerns about the public-health impacts of wireless technology—especially its effect on children—is Norbert Hankin, an environmental scientist in the EPA's Office of Radiation and Indoor Air. Hankin, a radio wave researcher, wrote of his concerns in an April 26, 2000, e-mail to Carlo's office:

> ". . . I suggest that another area of concern that should not be overlooked due to the potential impact on the quality of life of future adults (currently children) is the possible impact of wireless telecommunications technology and products on the learning ability of children.

> "The growing use of wireless communications by children and by schools, will result in prolonged (possibly several hours per day), long-term exposure (12 or more years of exposure in classrooms connected to computer networks by wireless telecommunications) of developing children to low-intensity pulse modulated radiofrequency (RF) radiation.
>
> "Recent studies involving short-term exposures have demonstrated that subtle effects on brain functions can be produced by low-intensity pulse modulated radiofrequency (RF) radiation. Some research involving rodents has shown adverse effects on short-term and long-term memory. The concern is that if such effects may occur in young children, then even slight impairment of learning ability over years of education may negatively affect the quality of life that could be achieved by these individuals, when adults. The potential effect on learning of exposure from telecommunication devices used by children should be considered for study by the Radiation Protection Project."

All available evidence to date makes this much clear: Even more than adults, children are put at serious risk from the radiation emitted by mobile phones. They definitely should not use mobile phones—at least not without a headset. Pagers are a better option for children who need to stay in touch because pagers require less energy and are not placed against the head.

What is not clear is why the top executives of the mobile phone industry have chosen to disregard the red-flag data that was waved in their faces years ago by Dr. Om Gandhi. What is unconscionable is that the government regulators of the FDA—whose job it is to safeguard the public—have chosen to not intervene or even, at the very least, issue appropriate cautions and warnings. It is the job of government officials to make sure parents everywhere know all that there is to know, to make sure they do not unwittingly put their own children at risk.

CHAPTER FIFTEEN

FOLLOW-THE-SCIENCE: *CONFIRMING EVIDENCE*

SOMETIMES THE MOST consequential events take place in the most inconsequential places. This was the case in the summer of 2000, when an event of potential significance to millions of mobile telephone users took place in a windowless room, on the ground floor of a nameless and faceless brown brick building, in a suburban office park in Rockville, Maryland.

The building houses the FDA's Center for Devices and Radiological Health. Yet a visitor could drive around forever amid the cluster of low-rise, look-alike buildings and not see a sign that said this was where government officials make decisions that can affect the physical health of 100 million Americans—and the financial health of multibillion-dollar corporations whose products they regulate. There is one prominent sign: It indicates the rather ironic address of this building—9200 Corporate Boulevard.

But the FDA's biggest sign problem is not about office signs but warning signs. FDA officials seem to have trouble seeing them.

Consider, for example, the warning sign that was waved before official eyes in that FDA building on August 1 and 2, 2000. It was a warning sign that became a focus of the two-day discussion: A distinguished scientist from the prestigious Washington University in St. Louis, Dr. Joseph Roti Roti, had produced findings that seemed to confirm those of Wireless Technology Research (WTR). Using different methods and systems, his research showed that mouse cells exposed to radiation at wireless phone frequencies did indeed develop micronuclei.

What made the finding all the more credible was the fact that Dr. Roti Roti, who is highly respected in his field, is hardly an anti-industry antagonist. He is a prominent scientist who does his work under sizable grants from Motorola Inc.

The importance of this FDA-hosted conference becomes clear only by knowing its political and scientific origins. This gathering of experts in the study of cellular telephone radiation was the first public event of a new, rather cozy partnership arrangement between the government and the industry—a formal Cooperative Research and Development Agreement (always written in memos from the government, industry, and science as "CRADA" and always pronounced as "the CRAY-da"). It was the same arrangement the industry had rejected in 1993 as a conflict of interest. The research mission outlined in this government–industry cooperative research agreement was a limited effort that came down to this: Go forward by backtracking. Follow up on the red-flag findings of the WTR's research. The industry had been raising doubts about the validity and importance of these WTR findings by emphasizing that the findings had never been replicated or otherwise confirmed by any other scientists. The findings of genetic damage in the form of micronuclei development in human blood cells that had been exposed to mobile phone frequency radiation were done for the WTR program by Dr. Ray Tice and Dr. Graham Hook, two others at the conference.

Truth is, Roti Roti's confirming findings came as no surprise to Tice and Hook; what did come as a pleasant surprise, though, was Roti Roti's willingness to go public at that meeting with his results—for Roti Roti had called Hook months earlier to say that

Roti Roti also had found a significant increase in micronuclei in blood cells exposed to radiation. The call came after yet another round of peer review that had been set up by Carlo. In mid-1999, the Harvard-based peer-review panel had come back with one concern about the studies of Tice and Hook. The peer reviewers wondered if it was possible that heat had built up during the radiation exposure of the test tubes, and if it was an artificial buildup of heat rather than radiation that caused the genetic damage. Because of that concern, Carlo arranged for a peer-review meeting, held at The George Washington University Club in Washington, where experts could review the procedures and perhaps suggest additional steps. Tice and Hook had been confident that they were aware of the danger of heat buildup and had monitored it carefully throughout the experiment to assure it did not become a factor. Later, Roti Roti called Hook to compare notes. He too had been watchful about guarding against heat buildup. At one point Roti Roti even proposed that he, Tice, and Hook jointly publish their findings. But a joint publication between a scientist tied to Motorola and scientists tied to Carlo—while an intriguing concept—never happened.

The industry, in its major public-relations effort to diminish the WTR findings, had skillfully succeeded in helping a less-than-vigorous news media overlook the fact that the WTR studies had been peer-reviewed to an extraordinary extent. Indeed, in an interviewer's conversation with journalists who have written extensively about this subject and about Carlo in particular, it was revealing that a number of the reporters, after having been "spun" by industry officials, were of the opinion that Carlo's WTR studies had never been peer-reviewed. That "spin" suggested to the reporters that the WTR findings might have been flawed. They were surprised when told by an interviewer for this book that the record is replete with instances in which the peer-review panel, under the guidance of the Harvard-based center, and other peer-review experts had been involved in the process from start to finish, contributing constructive suggestions every step of the way.

The official position of the cell phone industry and its junior research partner, the government, was that there still needed to be independent, outside, confirming evidence as a critical step toward determining whether there really was a public-health risk that warrants further government action, either through official intervention or even just warning-labels cautioning consumers. Roti Roti's study seemed to provide an important part of that—even before the ink was dry on the industry–government research pact.

A PEEK INSIDE THE FDA

Now, on August 1 and 2, 2000, in a modest-sized room in the FDA's Center for Devices and Radiological Health, two long tables had been placed in a V-formation so that various experts could face each other and still see a large projector screen at the front of the room. At the table to the left of the projector sat the government scientists and supervisors; at the table to the right sat some of the most famous scientists in the research of cell phone radiation. They had been invited to participate. Dr. George Carlo was not invited to this industry–government panel and did not attend. Sitting in the rear of the room, watching but not participating, were a mere 15 people who came to observe from the mostly vacant public section. A few of these were newspeople from the trade publications. A few were company people; CTIA vice president Jo-Anne Basile was there and sitting alongside her, whispering to her now and then, was Dr. Mays Swicord. For years Swicord had been the FDA's lead scientist and a major problem for industry leaders; later he had switched sides to become a top scientist on Motorola's payroll—and his most notable consistency in both capacities was that he had remained a critic of Carlo's efforts.

During the most significant—and scientifically stunning—portion of the conference, all eyes were focused upon a scientist dressed unpretentiously in a red-checkered sport shirt (in contrast to all the others, who wore shirts with ties), Dr. Roti Roti. He was sitting alongside Drs. Tice and Hook—which was fitting, because Roti Roti was outlining findings that appeared to confirm the WTR's studies by Tice and Hook that had so shaken the industry a year earlier.

Roti Roti told the scientific conference that he too had found that micronuclei had developed in blood cells exposed to cell phone radiation. Roti Roti's findings had been done with a similar though different exposure system, and similar though different methods—yet they seemed to confirm Tice and Hook's controversial findings that had been challenged so vigorously by the industry. "Our results are not a certain confirmation of theirs, but a possible confirmation," Roti Roti said, in an interview.

Roti Roti used radio frequency (RF) exposure systems at specific absorption rate levels of 1, 3, and 5 watts per kilogram (W/kg). He reported finding that at 5 W/kg the development of micronuclei occurred, and that the occurrence was statistically significant. His repeated experiments had obtained positive findings, but he said he had also gotten a negative finding in the same study. When the WTR studies had found the development of micronuclei at 5 and 10 W/kg the industry had mounted an all-out drive to discredit Carlo and the findings. Indeed, the industry sought to cast aspersions upon Carlo's findings by seeking to discredit him. The industry could not similarly discredit Roti Roti.

Roti Roti's findings became a new focus of the conference. The prime topic of discussion among the scientists from inside and outside government was no longer whether the development of genetic damage to human blood cells in the form of micronuclei was real or a significant warning of health risk. Now the debate was over whether the micronuclei had developed due to a thermal effect—heating—of the blood cells in the prolonged exposure. The question became: Was the biological change due to a laboratory anomaly or mobile phone frequency radiation? If heat alone was the cause, that would lead scientists to discount the possibility that the radiation could damage humans. Tice and Hook said they had developed their procedures to make sure they did not create a heating effect that could negate their findings. They had used a specially designed exposure system, and said they had constantly monitored the temperatures for verification.

Tice and Hook voiced confidence that their findings were the effects of radiation and not heating. "I think our experiments were well-controlled for heating effects," Hook said in an interview during a conference recess. Hook had sat next to Roti Roti and listened with more than just dispassionate interest as evidence was presented by an independent scientist that could only be seen as confirming the

WTR findings that others had viewed so skeptically. "This is great for us," Hook said. "His study used similar dose levels [of radiation] to ours. Now two laboratories, using two different systems, have gotten the same basic result."

Roti Roti, a highly respected researcher, was caught in a bit of the bind that is so familiar to many scientists. He needed to be true to his research, and yet grants from Motorola remained very much a part of his efforts to follow the science. He offered a similar but slightly different assessment than that of Tice and Hook. "They say it is less likely that there was a heating effect that caused this development of micronuclei," Roti Roti in an interview outside the conference room. "I say it is not certain." But he emphasized that he too monitored the temperatures to ascertain that no laboratory heating effect was artificially influencing his positive findings of micronuclei development.

Then there was the interpretation offered by another well-known expert on the panel, Dr. C. K. Chou, who had done some of the pioneering work on this problem in conjunction with the WTR's people and had later gone to work for the industry, becoming director of Motorola's Corporate RF Dosimetry Laboratory in Ft. Lauderdale, Florida. It was against that background that Chou said in an interview that while he found Roti Roti's findings important, he felt there must be more research to rule out the possibility that the genetic effects could have been caused by the buildup of heat during the exposure process. Still, while Chou works for industry these days, he stopped short of walking down the industry's see-no-problem path. "Now we're all like people looking at an elephant," Chou said in an interview. "One sees just the trunk, another just the tail. I must say I don't know what the animal looks like. We want to make sure this is not any threat to health."

It was late in the afternoon of August 2 when the conference ended, and Roti Roti, Chou, Tice, and Hook were conferring about calling a taxi that they would share on the ride to Washington's Ronald Reagan National Airport, where they would go their separate ways and return to their separate research. At day's end this much was clear: The scientists in the room all understood the confirming importance of Roti Roti's studies.

No longer was this just a case of George Carlo's WTR findings against the world. No longer could it be said that the science had produced no evidence that radiation from mobile phones can cause biological change in human blood cells—one definition of a risk to human health.

But those who relied upon the government regulators and watchdogs to communicate a new message of caution to the people who pay their salaries would be sadly disappointed—as they would discover just one week later.

CHAPTER SIXTEEN

MYOPIC WATCHDOGS

BY THE SUMMER OF 2000, cell phones and cancer was a topic that was on the minds of millions around the world. So it seemed fortuitous that just seven days after Dr. Joseph Roti Roti outlined his results that confirmed the original red-flag findings of the Carlo research team, the topic of cell phones and cancer was again beamed out to a national—in fact, global—audience on CNN.

Once again, CNN's *Larry King Live* was taking on the issue of cell phones and cancer. Once again, King was opening his show with a brain-cancer victim's lawsuit against the cell phone industry. But this time he scheduled as his final guest of the evening Dr. David Feigal, the head of the FDA's Center for Devices and Radiological Health—the watchdog agency that had just hosted the conference where Roti Roti presented his findings that cell phone radiation causes genetic damage to human blood cells.

Surely, you must be thinking, this was the time that the FDA would perform its obligation to warn the public of the new findings and the increased need to act with caution and protect ourselves as we communicate in the wireless age. You are right to be thinking that. But sadly, you are also wrong. The FDA chose instead to strike a public posture that was virtually indistinguishable from the industry it was supposed to be watching and regulating.

Larry King Live opened with a heart-tugging guest—a Baltimore physician with a brain tumor in the precise location where he'd always held his cell phone against his head. He was suing the manufacturer of his mobile phone, Motorola, and his phone service provider, Verizon, for $800 million. Then King read the industry's two written statements, which had but one theme: No health risk.

"We invited Motorola Inc. to appear on tonight's show," said King. "They declined. They did send us a statement. It says, in part, '. . . Over the years, scientific expert panels, standard-setting organizations, and other authoritative bodies around the world have not wavered from the longstanding conclusion that the low-power radio signals from wireless phones pose no known health risk.'

". . . The other company . . . Verizon, also declined to appear, but we have this statement from Nancy Stark, a spokesperson for Verizon Wireless: 'We can't comment on matters in litigation. On the general subject, I would refer you to the FDA's recent Consumer Update on Mobile Phones, which concludes that the available scientific evidence does not demonstrate any adverse health effects.'"

Then came the parade of scientists. Featured were two scientists, one who has testified as an expert witness for the industry and another who does consulting for the industry—facts of which King's producers apparently were unaware of, since the connections were never communicated to viewers. These scientists said they see no health-risk problem from wireless phones, as in this exchange between Larry King and Dr. John Moulder, of the Medical College of Wisconsin, who has been an expert witness in litigation:

> **KING** Dr. Moulder, the British government recommends discouraging kids from using cell phones, and Dr. Carlo, who headed a cell phone industry's six-year cell phone study, says evidence indicates kids

> could be a special risk. Should we refrain from kids using them?
>
> **MOULDER** Well, from a biological health standpoint, there's no particular reason why kids should be at any greater hazard than adults.

The show also booked a medical doctor who writes for *Time* magazine and had consistently provided analysis indistinguishable from that of the cell phone industry's press agents. In fact, the King show booked only one so-called expert who said cell phones *could* cause cancer, a chronic-disease epidemiologist who appeared to be a generation older than the other panelists, and who was familiar with studies that pre-dated the wireless era but seemed unfamiliar with the most current findings.

A viewer might have wondered why the deck seemed stacked in favor of the industry's position, but the FDA—the people's watchdogs and protectors—would certainly appear any minute now to add the missing perspective. And viewers were surely in luck, for King's FDA guest was the official whose center had just a week earlier hosted the groundbreaking conference where independent evidence confirming the Carlo team's findings had been the prime topic. Surely vital news would soon be shared, vital public precautions would soon be aired.

"And now the government's side of things," King said, late into his show. "Joining us, Dr. David Feigal. He's director of the FDA Center for Devices and Radiological Health, based in Washington. As of this day, Dr. Feigal, what's the FDA's position on the cell phone?"

Dr. Feigal replied, "We've reviewed the studies that have been discussed by your panelists in the previous segment, and it's our conclusion that at this time there is no reason to conclude that there are health risks posed by cell phones to consumers."

Remember: Feigal was the head of the FDA center that just a week earlier hosted the conference where the world renowned scientist and frequent Motorola researcher, Dr. Joseph Roti Roti, discussed his findings that, like those of the WTR, found mobile phone frequency radiation caused micronuclei in human blood. Yet, inexplicably,

Feigal, in his two long segments on the King show, never told the public about Roti Roti's confirming findings about micronuclei. Instead, he made a statement that sounded as if it could have been written by the cell phone industry itself. (Rewind and replay: Motorola had said, "Wireless phones pose no known health risk." Now the FDA's Feigal said: "There is no reason to conclude that there are health risks posed by cell phones to consumers.")

Now remember what world cancer experts have said about micronuclei in human blood cells: After the 1986 nuclear disaster in Chernobyl, international experts used micronuclei testing as a vital tool for diagnosing which children were at high risk for cancer and needed preventative treatment—a vital heads-up that saved lives. Also, cancer experts around the world and at the most respected institutions in the United States have used micronuclei tests in the same way. And in the August 2000 issue of the *Journal of the National Cancer Institute*—which is also a part of the federal government—two doctors had written that the presence of excessive micronuclei indicates that cells can no longer repair broken DNA, a deficiency that will "likely lead to the development of cancer."

You might expect that a publicly minded government official, after two new red-flag findings of micronuclei in human blood cells that were exposed to mobile phone radiation, would be motivated to pass along to the people some cautionary guidance. You might expect that, at least, the official would offer some interim steps that his agency might recommend so that mobile phone users can protect themselves until the FDA could make a final determination of safety or danger—perhaps just that people should use a headset to keep the antennas away from their bodies.

But no. Instead the FDA's Feigal seemed to go out of his way to make viewers think that any precaution such as using a headset is just something that *might* calm the worrywarts in our midst.

Feigal put the onus entirely on the consumer, giving no hint that the need for safety precautions had any scientific validity, let alone government support. "If a consumer is concerned and wishes to reduce their risk of exposure," the FDA official said, "then they can do the kind of things that were discussed [earlier in the show]: limit the duration of calls, use a headset, try and do things that minimize the amount of time that an antenna is close to the head. But this again is a precautionary thing by someone who's concerned while waiting for the answers to come in."

King did try to press, but the people's myopic watchdog was operating in a see-no-danger, hear-no-danger, speak-no-danger mode. Which may explain why, throughout the show, the watchdog never barked.

> **KING** Some critics have said that the FDA, the government has not acted aggressively enough, that this happened with tobacco as well, that you're sort of letting the companies do the studies, you're not spending enough, you're not as involved in the hunt. How do you respond?
>
> **FEIGAL** Well, the tobacco is an interesting comparison. One of the things that was discussed in the last couple of segments is what are the biological effects of radiation from cell phones. There has been no problem demonstrating biological effects of tar. That was actually one of the first human cancers ever observed in chimney sweeps, and in any number of animal models you can easily produce tumors and other carcinogenic effects that have been very difficult to demonstrate for this kind of product.

The government's watchdog was not about to initiate any warning bark on any facet of the problem. For example, it fell to King to bring up the fact that the British Stewart Commission has recommended that children be prohibited from using mobile phones. Unfortunately, King apparently had not been armed with an arsenal of those ads for Mickey Mouse, Minnie Mouse, Donald Duck, and Goofy cell phones marketed to appeal to children—or with Dr. Om Gandhi's calculations for the huge increase in the way radiation penetrates children's heads as compared to adult heads. And most unfortunately, the FDA official was not about to venture a vital word of caution to parents who are being lured each day by those Disney phone ads.

> **KING** How about children and usage? We discussed that earlier and Britain suggests not. What do you think?
>
> **FEIGAL** I think the answer given by your previous panel is the current assessment, that it's a precaution that they have—that they have taken, not because there have been new studies on children but by extrapolating from potential concerns. They have identified use by children as an area to be particularly careful, because they potentially can have the longest exposures.

Feigal noted at one point that a number of federal agencies do have responsibilities that touch on different aspects of the wireless phone issue, "but because of our expertise in health and risk assessment for health, it falls to us." At the end of his show, King asked, "And so you are, the FDA is saying to the public what? . . . The viewer now—the viewer now has a cell phone. You are saying to him what?"

"We're saying two things," said the FDA official. "One is that we have reviewed independently the currently available data and we do not see a health risk from the current data. However, if someone wishes to take precautions, they should limit the duration of their calls. They should do—take the measures that move the antenna away from their head, including hands-free sets and other types of precautions, as a precaution while they wait for the answer to come in."

The FDA was in the midst of what can only be described as a hell-bent retreat. Back in July 1993 the FDA's Dr. Elizabeth Jacobson sternly took Tom Wheeler to task back on July 19, 1993, after the CTIA's top lobbyist had held a press conference saying that cellular phones were safe. She wrote to him, "Our position, as we have stated it before, is this: Although there is no direct evidence linking cellular phones with harmful effects in humans, a few animal studies suggest that such effects could exist. It is simply too soon to assume that cellular phones are perfectly safe, or that they are hazardous—either assumption would be premature. This is precisely why additional research is needed." And since that time, there have been no findings refuting the studies that had so concerned the FDA in 1993.

Indeed, evidence has begun to pile up to the effect that wireless phone radiation can cause biological effects in human blood—genetic damage that is clearly detrimental to human health.

Also, the FDA, whose top official once wrote of "Our expertise in health and risk assessment for health," took no public action when mobile phones decorated with Disney cartoon characters were marketed to unsuspecting parents who assumed these products were ideal safety-first gifts for their children. The FDA never brought to public attention the data that showed children face a much greater risk than adults from radiation penetration from cellular phones.

Three mid-level FDA officials who have been in charge of mobile phone health policy all declined through a spokesperson to be interviewed for this book unless they were given in advance the questions they would be asked. When co-author Martin Schram placed telephone calls to the offices of Dr. Elizabeth Jacobson, Dr. David Feigal, and Dr. Russell Owens, his calls were returned only by FDA spokesperson Sharon Snider. She said that these three public officials wanted to see a list of questions before they would indicate whether or not they would be willing to be interviewed for this book. Schram responded by telling the FDA spokesperson he had never been required to provide questions in advance in more than three decades of interviewing government officials at the highest levels, that to provide questions in advance would be journalistically improper, and that while officials in dictatorships might require questions in advance, public servants in a democracy have a responsibility to answer to the public they serve—especially in matters of public health and safety. But on two separate occasions, FDA spokesperson Snider replied that these three officials would not even consider being interviewed unless they could know in advance the questions they would be asked.

WATCHDOGS IN RETREAT

Back on March 13, 1997, in a letter to Wireless Technology Research (WTR)'s Carlo, Dr. Jacobson of the FDA outlined the government's Interagency Working Group's view on the priorities by which it believed cell phone researchers should be guided. Two months later the FDA forwarded a copy of that letter to Rep. Edward Markey as evidence of the FDA's involvement in the scientific research that would be addressing the wireless phone health effect problems.

Dr. Jacobson's list contained seven major and comprehensive research priorities:

"Chronic (lifetime) animal exposures should be given highest priority.

"Chronic animal exposures should be performed both with and without the application of chemical initiating agents to investigate tumor promotion in addition to tumorigenesis.

"Identification of potential risks should include endpoints other than brain cancer (e.g., ocular effects of radio-frequency radiation exposure).

"Replication of prior studies demonstrating positive biological effects work is needed. A careful replication of the Chou and Guy study (*Bioelectromagnetics* 13:469–496, 1992) which suggests that chronic exposure of rats to microwaves is associated with an increase in tumors . . .

"Genetic toxicology studies should focus on single cell gel studies of DNA strand breakage and on induction of micronuclei. (These are the only direct genotoxic effects suggested at this time.) The need to replicate the Lai and Singh experiments . . .

"Epidemiology studies focused on approaches optimized for hazard identification are warranted . . .

"Indeed we believe that continuing post marketing surveillance is important in ensuring the safety of wireless technologies."

In 2000, the FDA's three-year old list of priorities stood as a research equivalent of a ghost town—abandoned and serving only as a curiosity. For in 2000, the FDA and the CTIA signed their formal Cooperative Research and Development Agreement (CRADA). It formalized the parameters of the research on cell phone radiation that would now be done: just follow up on the WTR's micronuclei research by trying to repeat those studies, rather than go on from there to do the rest of the research that was needed all along and which has never been done. This was a page not so much from the theories of General Lee or Jackson or Grant as from the playbook of Coach Dean Smith, the North Carolina basketball legend. It was the four-corner stall—an effort to buy time by slowing the game to a standstill. The only difference between that old college basketball ploy and the contemporary FDA's bureaucratic effort was that the NCAA recognized the flaw and installed a shot-clock that made such things illegal in basketball. But the government still sees the four-corner stall as business as usual.

Under the government–industry research agreement there would be no new animal studies, and none of the human studies that the government itself had once considered vital. And there would be no tracking and monitoring of the long-term health of people who use mobile phones.

The FDA's retreat in research goals is particularly hard to understand given the fact that their May 5, 1997, letter to Rep. Markey, forwarding the long list of priorities they had given Carlo, also stated: "The Agency has no information which would lead us to believe that the Federal Government and the American consumers will not be able to rely on the results of the WTR research."

COMPARISON OF RESEARCH PRIORITIES 1997 VS. 2000

SCIENTIFIC WORK ADVOCATED BY FDA IN 1997 LETTER TO WTR THAT WAS LATER CONVEYED TO CONGRESS

FDA Priority Suggested to WTR	Research Covered in CRADA
Chronic lifetime animal studies	NO
Chronic exposures of animals with initiating agents	NO

Chronic exposures of animals without initiating agents	NO
Studies of risks other than brain cancer	NO
Replication of prior animal study showing cancer	NO
Genetic toxicology studies	NO, but will critique WTR findings
Epidemiology studies	NO, but will consider feasibility
Post-market surveillance	NO

• • •

Under the provisions of the research agreement known as the CRADA, the government agreed that the cell phone industry would pay for all the research studies—and what the industry would get for its money was the right to make all final decisions about which studies would be done and which scientists would do them. The industry and its designated scientists would also have the final say on whether and when the findings would be published. The firewall of financial independence between the research coordinators and the industry—an independence that the GAO had required in 1995—no longer existed. The government regulatory agency limited its role to giving advice and being advised. But, the FDA was also giving one more thing: its de facto government seal of approval for whatever results would be brought forth by this so-called partnership research effort.

In the executive branch and in the Congress, the watchdogs that were supposed to protect the public interest suffered from either a bureaucratic myopia or political laryngitis—or both. Either way, these watchdogs never barked. They have failed to warn the people they are paid to serve.

PART FOUR

CHAPTER SEVENTEEN

FOLLOW-THE-SCIENCE: *PIECING TOGETHER THE CANCER PUZZLE*

SCIENTIFIC FINDINGS ARE like pieces of a puzzle. Individually, they may not seem to show anything clearly. But by trying to fit the pieces together, it is possible to see if they form a big, coherent picture.

In the puzzle of cell phone radiation research, the pieces of scientific evidence we have now do fit together. Although many pieces are still missing, those that are in place indicate a big picture of cancer and health risk. The picture is alarming, because even if the risk eventually proves to be small, it will still be real—and that means millions of people around the world will develop cancer or other health problems due to using mobile phones.

Even more alarming, however, is that many in the industry, who are paid by the industry—and some who are paid by the public to oversee and regulate the industry—have persisted in talking publicly as if they cannot see the picture that is taking shape even as they speak.

In the study of public health, there is a well-known template that researchers use to put together individual scientific findings—like the pieces of a puzzle—to see if they show evidence of a public-health hazard. This template, known as the Koch-Henle Postulates, is a means of determining whether the findings indicate a true cause-and-effect process, from biological plausibility to exposure and dose–response. The postulates are:

1. If there is a biological explanation for the association derived from separate experiments that is consistent with what is known about the development of the disease, then the association is more likely to be causal. Scientists term this *biological plausibility*.

2. If several studies of people are showing the same finding while employing different methods and different investigators, the association that is being seen is more likely to be cause and effect, or causal. Scientists term this *consistency*.

3. If it is clear that the exposure precedes the development of the disease, then the association is more likely to be causal. Scientists term this *temporality*.

4. If the increase in risk is significant—more than a doubling in the risk or an increase that is statistically significant—the association is more likely to be causal. Scientists term this *significance*.

5. If the more severe the level of exposure, the higher the risk for the disease or the biological effect that is being studied, the association is more likely to be causal. Scientists term this *dose–response upward*.

6. If the absence of exposure corresponds to the absence of the disease, the more likely the association is to be causal. Scientists term this *dose–response downward*.

7. If there are similar findings in human, animal, and *in vitro* studies—in other words, if the same conclusions can be drawn from all—the more likely the association that is seen is causal. Scientists term this *concordance*.

Researchers use the Koch-Henle Postulates as a checklist. The greater the number of postulates that are met, the greater the likelihood that a hazard exists. For some of the more commonly recognized carcinogens, it has taken decades for the hazards to be judged as valid. For example, in the case of cigarette smoking, it took two decades of study and more than 100 years of consumer use to gather enough information that could be judged against the Koch-Henle standards to demonstrate the need for the U.S. Surgeon General's warning label on cigarette packs.

In the case of cellular telephones, consumers are fortunate that the health-hazard picture can be seen much sooner than that. Each of the red-flag findings about cell phone radiation provides a vital piece of information that fits into the overall cancer puzzle. A number of the other earlier studies, which on their own were inconclusive or seemed uninterpretable, now appear to fit into the puzzle as well. They clarify a troubling picture of cancer and health risk that is just now becoming clear.

Here is how the scientific pieces fit into the larger cancer puzzle:

- **Human Blood Studies:** These studies—by Drs. Ray Tice and Graham Hook, and most recently corroborated by Dr. Joseph Roti Roti—show genetic damage in the form of micronuclei in blood cells exposed to cell phone radiation. They provide evidence of the Koch-Henle postulate of *biological plausibility* for the development of the tumors following exposure to radio waves. Without some type of genetic damage, it is unlikely that radio waves would be able to cause cancer. Every direct mechanism that has been identified in the development of cancer involves genetic damage;

the linkage is so strong that if an absence of genetic damage had been proven in these studies, scientists would have considered that to be reason enough to conclude that cancer is not caused by cell phones. (Indeed, that is what scientists were justified in saying prior to 1999.) Scientific literature has repeatedly confirmed that brain cancer is clearly linked to chromosome damage; brain tumors have consistently been shown to have a variety of chromosomal abnormalities. The studies by Tice, Hook, and Roti Roti, consistently showed chromosomal damage in blood exposed to wireless phone radio waves.

- **Breakdown in the Blood Brain Barrier:** The findings of genetic damage by Tice, Hook, and Roti Roti now give new meaning and importance to Dr. Leif Salford's 1994 studies that showed a breakdown in the blood brain barrier of rats when they were exposed to radio waves. The blood brain barrier findings now fit into the overall cancer picture by providing a two-step explanation for how cancer could be caused by cell phone radiation. (The blood brain barrier filters the blood by not allowing dangerous chemicals to reach sensitive brain tissue.)

  Step One: A breakdown in the blood brain barrier filter would provide an avenue for chemical carcinogens in the bloodstream (from tobacco, pesticides, or air pollution, for example) to leak into the brain and reach sensitive brain tissue that would otherwise be protected. Those chemicals, upon reaching sensitive brain tissue, could break the DNA in the brain or cause other harm to reach those cells.

  Step Two: While a number of studies showed that cell phone radiation by itself does not appear to break DNA, the micronuclei findings of Tice, Hook, and Roti Roti suggest that DNA repair mechanisms in brain cells could be impaired by mobile phone radiation. (One reason micronuclei occur is that there has been a breakdown in the cell's ability to repair itself.) If the brain cells become unable to repair themselves, the process of chemically induced carcinogenesis—the creation of tumors—could begin.

This is further evidence of the Koch-Henle postulate of *biological plausibility* for cell phone radiation involvement in the development of brain cancer.

- **Studies of Tumors in People Who Use Cell Phones:** There have been four studies of tumors in people who use cellular phones—Dr. Ken Rothman's study of deaths among cell phone users, Joshua Muscat's two studies of brain cancer and acoustic neuroma, and Dr. Lennart Hardell's study of brain tumors. All four epidemiological studies, done by different investigators who used different methods, show some evidence of an increased risk of tumors associated with the use of cellular phones. This is evidence of the Koch-Henle postulate of *consistency*.

  All four epidemiological studies provide some assurance in the methods used by the investigators that the people studied had used cellular telephones before they were clinically diagnosed as having tumors. This is evidence of the Koch-Henle postulate of *temporality*.

  All four epidemiological studies showed increases in risk of developing brain tumors. Muscat's study of cell phone users showed a doubling of the risk of developing neuro-epithelial tumors. (The result was statistically significant.) Hardell's study showed that among cell phone users, tumors were twice as likely to occur in areas of the brain at the side where the user normally held the phone. (This result was also statistically significant.) Rothman's study showed that users of handheld cell phones have more than twice the risk of dying from brain cancer than do car phone users—whose antennas are mounted on the body of the car, far removed from the users' heads. (That finding was not statistically significant.) Muscat's study of acoustic neuroma indicates that cell phone users have a 50-percent increase in risk of developing tumors of the auditory nerve. (This finding was statistically significant only when correlated with the years of cell phone usage by the patient.) These findings are evidence of the Koch-Henle postulate of *significance*.

- **Studies of Cell Phone Radiation Dosage and Response:** In Dr. Michael Repacholi's study of mice, the risk of lymphoma increased significantly with the number of months that mice were exposed to the radio waves.

  In the work by Tice, Hook, and Roti Roti, the risks of genetic damage as measured by the formation of micronuclei increased as the amount of radiation increased.

  In the three epidemiological studies—two by Muscat and one by Hardell—that were able to estimate radiation exposure to specific parts of the brain, the risk of tumors was greater in the areas of the brain near where the cell phone was held.

  These findings are all evidence of the Koch-Henle postulate known as *dose-response upward.* (In cell phones, minutes of phone usage are not a reliable indication of dosage, because the distance of the telephone from a base station during the call and any physical barriers to the signal are the most important factors in the amount of radiation the phone antenna emits during the call.)

  The Hardell epidemiological study showed that patients with tumors in areas of the brain that could not be reached by radiation from a cell phone antenna were likely not to have been cell phone users. Similarly, in Muscat's study, when all brain-tumor patients were included in his analysis—those with tumors that were outside the range of radiation from the cellular phone antenna and those whose tumors were within that range—there was no increase in the risk of brain cancer. This is evidence of the Koch-Henle postulate that is called *dose-response downward*—which simply means that if there is no chance that cell phone radiation dosage could have been received, chances are the tumor was caused by something else.

- **Agreement of Findings From *In Vitro* and *In Vivo* Studies:** The test-tube studies by Tice and Hook; the mouse study by Repacholi; and the four epidemiological studies by Rothman, Muscat, and Hardell are all in agreement in that they suggest an increase in the risk of cancer among people who use mobile phones. This is evidence of the Koch-Henle postulate of *concordance*.

. . . AND THE LARGEST PIECE OF THE PUZZLE

As the officials of the government, officials of the industry, and just plain unofficial people try to fit together jigsaw pieces to see whether mobile phones indeed pose a cancer risk, the cancer experts themselves have provided what is by far the biggest and most revealing piece of the puzzle. Writing in the U.S. government's own *Journal of the National Cancer Institute,* and other prestigious professional publications, these experts have made it clear that, if there are findings that micronuclei develop in blood cells exposed to mobile phone radiation, that is in itself evidence of a cancer risk. The risk is so persuasive, the experts have written, that preventative treatment should be given in order to best protect those people whose levels of micronuclei have increased.

THE BIG PICTURE

The pieces of the cell phone puzzle do indeed fit together to form the beginnings of a picture that researchers, regulators, and mobile phone users can all see for themselves. Many pieces are still missing. But enough pieces are already in place to see that there are legitimate reasons to be concerned about the health of people who use wireless phones.

Most alarming to public health scientists should be the fact that all seven of the Koch-Henle postulates have been met within the first decade of widespread mobile phone usage.

The big picture is becoming disturbingly clear: There is a definite risk that the radiation plume that emanates from a cell phone antenna can cause cancer and other health problems. It is a risk that affects hundreds of millions of people around the world. It is a risk that must be seen and understood by all who use cell phones so they can take all the appropriate and available steps to protect themselves—and especially to protect young children whose skulls are still growing and who are the most vulnerable to the risks of radiation.

CHAPTER EIGHTEEN

SAFETY FIRST: *HEALTH RECOMMENDATIONS*

As the big picture becomes clear and we see that radiation from mobile phones poses a real cancer and health risk, it also becomes clear that there are basic recommendations that now demand the urgent attention of all who use, make, research, or regulate cell phones.

Mobile telephones are a fact of life and a fixture in the lifestyles of more than half a billion people around the world. That only makes it all the more vital that we understand and follow the recommendations by which all who use mobile phones can minimize their health risk, and especially can protect our children. Here are some basic suggestions for mobile phone users, manufacturers, and science and medical researchers.

RECOMMENDATIONS FOR CONSUMERS

To avoid radiation exposure and minimize health risks when using wireless phones:

1. The best advice is to keep the antenna away from your body by using a phone with a headset or earpiece. Another option is a phone with speakerphone capability.

2. If you must use your phone without a headset, be sure the antenna is fully extended during the phone's use. Radiation plumes are emitted mainly from the mid-length portion of the antenna; when the antenna is recessed inside the phone, the entire phone functions as the antenna—and the radiation is emitted from the entire phone into a much wider area of your head, jaw, and hand.

3. Children under the age of ten should not use wireless devices of any type; for children over the age of ten, pagers are preferable to wireless phones because pagers are not put up to the head and they can be used away from the body.

4. When the signal strength is low, do not use your phone. The reason: The lower the signal strength, the harder the instrument has to work to carry the call—and the greater the radiation that is emitted from the antenna.

5. Emerging studies, and common sense, make clear that handheld phones should not be used while driving a vehicle.

A Few Words of Caution for Consumers

The public is bombarded with waves of claims that are made at times by individuals who are well-meaning but not well-informed—and at other times by special interests who really want to sell a product. For example, there is no scientific basis for recommendations that have been made by some groups to limit phone use as a means of minimizing the risk of health effects. It is not possible to determine scientifically the difference in radiation exposure from one ten-minute call and ten one-minute calls. The total number of minutes is the same, but the pattern and amount of radiation could be very different. Also, the amount of radiation emitted by a mobile phone depends on the distance of the phone from a base station; the further the distance, the harder the phone has to work and the greater the radiation. Finally, the greatest amount of radiation emitted by a phone is during dialing and ringing. People who keep their phones on their belts or in their pockets should move the phones away from their bodies when the phones are ringing. (The amounts of radiation in a single call can vary by factors of ten to 100 depending on all of these variables.)

Consumers also need to be cautious about unverified claims that seem to have scientific backing. For example: The media recently carried an account published in Britain's *Which?* magazine that said that a group called the Consumers' Association (with which the magazine is affiliated) had shown in tests that some cell phone headsets actually cause more radiation to go to the brain than the phones themselves. But the claim is unsubstantiated by any scientific evidence, and has been refuted by a number of studies by recognized researchers using established scientific methods. The only conclusion that can be drawn from existing scientific evidence is that headsets are the best option for mobile phone users to minimize exposure to wireless phone radiation.

Also, a number of devices on the market claim to eliminate the effects of antenna radiation and are being marketed as alternatives to using headsets or speakerphones. These products need to be tested to see if they will really protect consumers—a caution expressed by Great Britain's Stewart Commission. They recommended that their government set in place "a national system which enables independent testing of shielding devices and hands-free kits . . . which enable clear information to be given about the effectiveness of such devices. A kite mark or equivalent should be introduced to demonstrate conformity with the testing standard." In the United States, the FDA has been silent on the matter.

RECOMMENDATIONS FOR THE MOBILE PHONE INDUSTRY

To enhance consumer protection:

1. Phones should be redesigned to minimize radiation exposure to consumers—antennas that extend out at an angle, away from the head, or that carry the radiation outward should be developed.

2. Headsets and other accessories that minimize radiation exposure should be redesigned so they are more durable and can be conveniently used.

3. Consumers should be given complete information about health risks and solutions through brochures, product inserts, and Internet postings so they can make their own decisions about how much of the risk inherent to mobile telephone use they wish to assume.

4. Emerging and advancing phone technologies need to be premarket tested for biological effects so dangerous products do not make it to the market,

5. Post-market surveillance is necessary for all phone users—surveys of analog and digital phone users to see if they experience any adverse health effects, and databases should be maintained where people can report any health effects they have experienced due to their phones.

Recommendations for Scientific, Medical, and Public Health Officials

To help consumers:

1. Science, medicine, and government must move immediately and aggressively with the goal of minimizing the impact of radio waves on adults, children, and pregnant women.

2. One federal agency must be designated as the lead agency for protecting people who use wireless communications devices, rather than having the responsibility remain undefined and shared among multiple agencies including the FDA, FCC, EPA, and others.

3. A genuine safety standard needs to be established to serve as the basis for future regulatory decisions. Since the specific absorp-

tion rate alone does not measure biological effects on humans, it does not serve the safety needs of consumers.

RECOMMENDATIONS CONCERNING HEART PACEMAKERS

To enable patients to avoid interference from mobile telephones:

1. Wireless phones should be kept a safe distance from pacemakers—6 inches has been scientifically determined as the proper separation distance for minimal risk. The greater the distance between a pacemaker and a wireless phone, the less the risk of electromagnetic interference.

2. Do not keep a cell phone over the pacemaker, such as in the breast pocket, when it is in the "On" position.

3. Pacemaker patients should use analog phones, rather than digital phones; analog phones have a lower risk of interference than digital phones.

To enable physicians to best safeguard their patients:

1. Since pacemakers are now being made with special filters that are resistant to interference from mobile phones, patients who are mobile phone users should be informed that they now have this new, safer option.

2. Physicians should be aware that patients who are most severely ill and require a pacemaker to keep their heart beating have a higher risk of clinically significant interference from mobile phones; physicians need to inform these patients of this danger.

3. Physicians should not conduct testing of wireless phones and pacemakers on their own, because readings in non-laboratory locations are likely to be unreliable without the controls used in rigorous clinical studies.

To insure that pacemaker and mobile phone manufacturers solve the interference problem:

1. Before new pacemaker models are marketed, they should be screened for susceptibility to interference from wireless phones.

2. Pacemaker and phone manufacturers should place on all packages and/or equipment clearly visible labels that provide warnings about possible pacemaker/phone interference.

3. Post-market surveillance of pacemaker patients must be done to check for possible future cases of interference from mobile phones.

RECOMMENDATIONS FOR INDUSTRY AND GOVERNMENT CONCERNING THE WIRELESS INTERNET

We need to recognize and learn from the mistakes we made when cellular phones were first introduced. The phones were sold to the public before there had been any premarket testing to determine whether they were safe or posed a potential health risk. Because the cell phones were not tested initially, by the time they were on the market, efforts to research the problem became intertwined with the forces of politics and profit. Consumer protection was not the highest priority.

As we enter the age of the wireless Internet, no one can say for sure whether or not the radio waves of the these new wireless products will prove harmless or harmful. But this much is known: The concern about mobile phones focuses on the near-field radiation that extends in a 2-to-3-inch plume from an antenna, and the radiation from the many wireless laptop and handheld computer products is just about the same. It would seem that these latter products should be safer because users don't hold their laptops and handheld computers against their heads. But no one has researched what the effect will be of a roomful of wireless products all being used simultaneously, with radio waves invisibly crisscrossing the space that is occupied by people. Will these passive occupants run a risk similar to nonsmokers in a room filled with smokers, who end up affected by passive smoke?

Thus, it is important that these new products must be formally tested under official regulatory control that includes specific premarket screening guidelines. There must also be post-market surveys of people who use the wireless Internet to see if health problems emerge that were not found in the premarket testing.

Epilogue

On a pleasantly warm February day, Wireless 2000 was in full swing in the industry's favorite convention city of New Orleans, and George Carlo was making one last appearance where he had once been the trusted science insider. He was there for just one purpose: to formally present the final report of his Wireless Technology Research program before the board of directors of the trade association. He didn't really want to be there. And it would be clear soon enough that Tom Wheeler and the CTIA chieftains didn't really want him there either.

Upon arriving at the designated meeting room in the convention center on the banks of the Mississippi River, Carlo and his two colleagues, Cindy Perno and Lisa Joson, were intercepted by a security force of two big, beefy men in plainclothes. Their job was clearly to make sure the man who seven years earlier had been Tom Wheeler's handpicked insider was now secured and watched every moment that he was inside the hall. The two plainclothesmen, all muscle and girth and no necks—the larger explained he was ex–Secret Service—escorted Carlo and his colleagues to a waiting room. They told Carlo to go inside and stationed themselves at the door, apparently to keep reporters—including a CBS camera crew and producer—from interviewing the man who once helped plan the CTIA's press conferences. Fifteen minutes later, Carlo was escorted to the boardroom. He entered to a reception of absolute silence and briefly presented his final report, which told the industry's top executives the bad news they already knew. Wheeler rose, extended his hand, and said, "Thank you, George." And Carlo left as he had arrived, under security escort, with the board of directors staring at him silently as he walked out of the room. The security guards remained at Carlo's side until he had left the hall and hailed a curbside taxi.

As the taxi took Carlo and his colleages to the airport, he reflected upon how dramatically things had changed during his seven-year relationship with the cell phone industry and Tom Wheeler.

Before I met with the CTIA board that day, I slipped into the exhibition hall unnoticed for a quick look at Wireless 2000. The glittering exhibits of wireless high-tech were far different from anything that was on display at the CTIA convention seven years earlier. This association, after all, was named the Cellular Telecommunications Industry Association, but by 2000 the technology had clearly moved far beyond anything cellular. Indeed, today's new phones are not really cellular at all—they no longer just relay conversations from one cell or station to the next. The new generation of phones are called personal communication systems because in addition to carrying voices, they transmit data via fax and e-mail and connect to the Internet. Under Wheeler's command, his trade association had overcome the limits of its own name and had vaulted into a brave new wireless world. The flashing, blinking exhibits of Wireless 2000 featured computer vendors and online services extolling the wonders of the wireless Internet. At the first industry trade show I attended in 1994, the exhibits had been all about the wonders of telephones. But today, cell phones are to the wireless world what eight-track tapes are to digital minidisks.

The convention program too had leaped into a new age under Tom Wheeler. Once, he had made sure his program would be the talk of the town by bringing in Tony Bennett to sing and former President George Bush to do a fireside chat. But, in 2000 he sent a new-millennium message, bringing in Microsoft's Bill Gates, and America Online's Steve Case, to deliver the cell phone association into the wireless age.

It occurred to me that for all I could see in the glittering and impressive exhibition of wireless technology, the real key to the future could be found in what couldn't be seen. For unseen but ever-present in that cavernous hall were rays of microwave radiation. The radiation rays were crisscrossing the floor space and the air space all the way up to the high ceiling. And they were intersecting, and perhaps penetrating, the bodies of the people in the room as they were gawking at the wonders of tomorrow.

These conventioneers had every reason to feel safe and no reason to feel at risk. After all, they had the ultimate in security protectors, people whose job it was to be ever-vigilant—the regulators in the executive branch of government, the watchdog committees in the Congress, the investigative journalists in the news media, the objective researchers in science, and the experts in the wireless industry. All the forces seemed in place to keep people safe. And yet no one was really on the case. The regulators were not regulating. The watchdogs were not watching. The investigative journalists were not investigating. The researchers were signing on with industry, losing their objectivity, and failing to follow their science. And the industry was basking in record profits.

The new technology is indeed wondrous and it is changing the way we live and work. But breakthroughs in technology must be balanced by a willingness to actively and responsibly safeguard the public health. The safety of hundreds of millions of people, as we enter tomorrow's wireless age, depends upon the willingness of a handful of leaders in government, industry, and science to put politics and profit aside—and do the right thing today.

GLOSSARY

Acoustic neuroma: A benign tumor of the acoustic nerve. Sometimes also called the auditory nerve, it controls hearing and runs from the ear to the brain stem.

Analog phone: A type of wireless telephone that sends signals in continuous waves similar to those in FM radios. The first cellular telephones were analog, but now analog phones have given way to digital phones.

Assay: A laboratory test where biological effects are measured and quantified.

Base station: A structure with antennas that relays the signal from one wireless phone to another and provides a link to the wired telephone network.

BEMS: Bioelectromagnetics Society. A professional association that organizes scientific meetings dealing with the effects of radio waves.

Benign tumor: A tumor or growth that is noncancerous.

Blood brain barrier: The filtering system in blood vessels in the skull that keeps dangerous chemicals from reaching sensitive brain tissue.

Carcinogenic: Something that causes cancer.

Case-control study: A type of human study in which people with the disease being studied (cases) are compared to people without the disease (controls).

Cellular phone: A type of wireless phone that operates in the 800-to-900-megahertz frequency band.

CDMA: Code Division Multiple Access. A type of digital phone signaling that allows for many calls to be carried on at one frequency by alternating signals sent in computer code.

Cohort study: A type of human study in which individuals exposed to a product (such as a cellular phone) are compared to people who are not exposed based on the occurrence of disease.

Colloquium: Scientific meeting where discussion among participants is encouraged.

Comet assay: Common name for Alkaline Single Cell Gel Microelectrophoresis assay, or SCG assay, designed to measure DNA damage. An electric current is run through DNA already broken, for example, by exposure to sunlight, alcohol, caffeine, and nicotine. Segments of broken DNA base pairs become positively or negatively charged, so the current makes the fragments move. Under the microscope, the trail of broken DNA in motion resembles the tail of a comet. The longer the tail of the comet, the more DNA damage. A brain cell with no DNA damage has no tail.

CRADA: Cooperative Research and Development Agreement. An arrangement between the federal government and industry to do research together.

CTIA: Cellular Telecommunications Industry Association. The trade group that represents the business interests of wireless telephone service providers such as AT&T Wireless, Verizon Wireless, MCI, and Sprint.

Defibrillator: A medical device implanted in cardiac patients that sends an electric current to the heart as a jump-start when the heart stops beating.

Digital phone: A type of cellular phone that sends signals that are pulsed.

Dosimetry: The science of measuring the amount of radiation that emanates from an antenna.

EMF: Electromagnetic field. The generic term for the radiation that comes from devices that push electric current or radio waves.

EMI: Electromagnetic interference. Interference with devices such as pacemakers is induced by the EMF.

EPA: U.S. Environmental Protection Agency.

Epidemiology: Human studies that examine disease in groups of people.

FCC: Federal Communications Commission. Government group responsible for overseeing the development and use of radio communications technology.

FDA: Food and Drug Administration. U.S. federal regulatory agency responsible for protecting consumers from radiation-emitting consumer products such as wireless telephones.

GAO: General Accounting Office. Investigative arm of the U.S. Congress.

GSM: Global System Mobile. The wireless phone system used in Europe and other parts of the world outside of North America.

Hertz: A term referring to the speed at which a radio wave travels. One hertz equals one cycle per second.

IAWG: U.S. government's Interagency Working Group established to provide ongoing input into the research program to assess the health impact of wireless technology.

In vitro study: A type of laboratory study done outside the body and in an artificial environment, with test tubes and petri dishes.

In vivo study: A type of laboratory study done with live animals.

Leukocyte: A white blood cell.

Leukemia: A cancer of the blood cells in bone marrow.

Lymphocyte: A type of white blood cell with immune system function.

Lymphoma: A cancer of the lymphoid tissue (found mainly in the lymph nodes and the spleen).

Malignant: Describing a tumor that is likely to grow continuously; cancerous.

Megahertz: Million hertz or million cycles per second. Refers to the speed at which a radio wave travels.

Micronuclei: Fragments of DNA that have defined membranes around them, that are an indication of genetic damage. Normal blood cells do not have micronuclei.

Mortality: Death.

Mutagen: Any agent which causes DNA damage or alters genes.

NIEHS: National Institute of Environmental Health Sciences, an agency of the federal government.

Pacemaker: A medical device implanted in cardiac patients that sends a continuous electric current to the heart to regulate beating.

PCS: Personal communication system. The new generation of wireless communication products.

Premarket testing: Government-required tests to determine the safety of consumer products prior to those products being made available to people.

Post-market surveillance: Tracking and monitoring of people who use consumer products such as wireless telephones to see whether they develop health problems associated with their use. This is a government requirement for most consumer products that have a risk of harming people.

PCIA: Personal Communications Industry Association. The trade group that represents the business interests of wireless communication-instruments service providers for example, pager companies such as MetroCall and PageNet.

PCS phone: A type of wireless phone that operates in the 1900-to-2200-megahertz frequency band.

PRB: Peer Review Board. A group coordinated through the Harvard School of Public Health to provide ongoing scientific critique of the research program studying the health impact of wireless technology.

RFR: Radio frequency radiation.

SAG: Scientific Advisory Group that was the precursor to the WTR. It advised the CTIA about health and safety issues, but later was established as an independent research entity, outside of the CTIA's control.

SAR: Specific absorption rate. The amount of energy from an antenna that passes through a biological tissue during a specified time period. It is measured in watts of power per kilogram of tissue.

SCG assay: Alkaline Single Cell Gel Microelectrophoresis assay, or SCG assay; see Comet assay.

TDMA: Time Division Multiple Access. A type of digital phone signaling that allows for many calls to be carried on at one frequency by alternating signals sent at different times.

TEM cell: Transverse electromagnetic cell. A closed chamber that delivers radio frequency radiation to test tubes and petri dishes.

TIA: Telecommunications Industry Association. The trade group that represents the business interests of wireless telephone manufacturers such as Motorola, Nokia, and Ericsson.

Toxicology: The scientific field that examines the impact of dangerous substances.

Wireless phone: A portable telephone with an attached antenna.

W/kg: Watts per kilogram.

WTR: Wireless Technology Research, L.L.C. The legal entity established by CTIA and the SAG to oversee the research and surveillance effort aimed at assessing the health impact of wireless technology. It was created in direct response to recommendations made by the General Accounting Office of the U.S. Congress.

Cast of Characters

Adey, Ross, M.D.; Veteran's Administration Medical Center, Loma Linda, Calif.; Motorola-funded researcher in effects of radiation exposure on mice.

Basile, Jo-Anne; Vice President, Cellular Telecommunications Industry Association (CTIA), trade association representing cellular telephone service providers.

Brusick, David, Ph.D.; leading international genetic toxicology expert.

Carrillo, Roger, M.D.; cardiologist at Mt. Sinai Hospital, Miami; performed cell phone–pacemaker research.

Chou, C. K., Ph.D.; City of Hope National Medical Center, Los Angeles; identified how Cray computer could be used in cell phone–brain damage research.

Cleary, Stephen, Ph.D.; early researcher in cell response to microwaves.

Collins, William; chairman of Metrocall, nationwide retailer of pagers and wireless phones; after hearing of Carlo's findings, immediately had his stores start giving written advice to every one of his customers about possible risks of cell phone use—one of the few companies to do so.

***Doll, Sir Richard, F.R.S., F.R.C.P.;** Emeritus Professor of Medicine, Oxford University, England; member, Peer Review Board.

Dreyer, Nancy, Ph.D.; Epidemiology Resources, Inc.; researcher whose study, with Dr. Ken Rothman, found six deaths attributed to brain cancer among 462 people who died in 1994.

Eger, Charles; attorney and Motorola employee.

Gabriel, Camelia, Ph.D.; top international expert in radio wave movement through biological tissue.

Gandhi, Om, Ph.D.; University of Utah; a pioneer scientist in cellular telephone research; his 1996 study defined the profoundly large differences in radiation exposure between children and adults using cell phones.

***Graham, John D., Ph.D.;** director, Center for Risk Analysis, School of Public Health, Harvard University; project director, Peer Review Board.

Guy, Arthur (Bill), Ph.D.; Emeritus Professor, University of Washington; member, Science Advisory Group; director, SAG Dosimetry Working Group.

†Hankin, Norbert N.; Environmental Scientist, Office of Radiation and Indoor Air, EPA; member, IAWG.

Hardell, Lennart, M.D., Ph.D.; Departrment of Oncology, Orebro Medical Center, Orebro, Sweden; his studies found that cell phone users' risk of developing a tumor in the area near the cell phone antenna was

2.4 times greater than that of developing a tumor in any other area of the brain, a statistically significant finding.

Hayes, David, M.D.; Mayo Clinic, Rochester, Minn.; researcher on pacemaker–cell phone interaction.

Hook, Graham, Ph.D.; Integrated Laboratory Systems, Research Triangle Park, N. C.; researcher who, with Dr. Ray Tice, found first evidence that cell phone radiation causes DNA breaks.

†Jacobson, Elizabeth Dr., Ph.D.; Deputy Director of Science, Center for Devices and Radiological Health, FDA; member, IAWG.

Lai, Henry, Ph.D.; University of Washington; reseacher, with Dr. N. P. Singh, on radiation exposure effects on rats.

Lukish, Thomas; CTIA Vice President for Health and Safety; left CTIA after supporting the full funding of WTR research on possible dangers of cell phones.

Markey, Edward, Rep. (D-Mass); chairman of U.S. House of Representatives Commerce Committee's Subcommittee on Telecommunications, Trade, and Consumer Protection until 1994, ranking minority member thereafter.

Maxfield, Elizabeth; former CTIA vice president overseeing the health and safety issue.

McRee, Donald, Ph.D.; coordinator of program under which Drs. Ray Tice and Graham Hook did their research, under a contract with WTR.

Meltz, Martin, Ph.D.; University of Texas; consultant to U.S. Air Force on radio wave issues; later, scientific consultant to CTIA on cell phone risk.

Mild, Kjell Hansson, Ph.D.; Swedish National Institute for Working Life; researcher on physical effects of using analog and digital cell phones.

Munro, Ian, Ph.D.; former official in Canadian Health Ministry; member, Science Advisory Group; director, SAG Toxicology Working Group.

Joshua Muscat, American Health Foundation; became lead researcher on the WTR's series of epidemological studies.

Najafi, Marjan; coordinator of WTR toxicology program.

Nesbit, Jeffrey; SAG liaison with FDA.

Nessen, Ron; CTIA Vice President for Communications; former press secretary to President Gerald Ford.

†Owen, Russell D., Ph.D.; Chief, Radiation Biology Branch, Center for Devices and Radiological Health, FDA; member, IAWG.

Peterson, Ronald; senior level scientist and engineer, Bell Laboratories; member SAG Dosimetry Working Group.

Polansky, Gerald; chairman of WTR Audit Committee.

Powell, Jody; head of Powell Tate, Washington, D.C., public-relations firm; former press secretary to President Jimmy Carter.

***Putnam, Susan W., Sc.D.;** Research Associate in Environmental Policy, Center for Risk Analysis, School of Public Health, Harvard University; project director, Peer Review Board.

Repacholi, Michael, Ph.D.; Royal Adelaide Hospital, South Australia; researcher who published first scientific evidence that cellular phones could cause cancer.

Rothman, Kenneth, DrPH; Epidemiology Research Institute; researcher whose study, with Dr. Nancy Dreyer, found six deaths attributed to brain cancer among the 462 people who died in 1994.

Roti Roti, Joseph, Ph.D.; Washington University, St. Louis, Mo.; radiation biologist; first scientist outside the WTR program to confirm findings of Drs. Graham Hook and Ray Tice.

Salford, Leif, M.D.; Swedish researcher who first found blood brain barrier breakdown connected with cell phone use.

***Sheppard, Asher R., Ph.D.;** consultant, Asher Sheppard Consulting; member, Peer Review Board.

Silva, Jeffrey; Washington, D.C.–based reporter for the industry publication, *Radio Communications Reports*

Singh, Narendrah (N. P.), Ph.D.; University of Washington; researcher, with Dr. Henry Lai, on radition exposure effects on rats.

Sivak, Andrew, Ph.D.; world-renowned expert on cancer; head of WTR's Expert Panel on Tumor Promotion.

Slesin, Louis, Ph.D.; editor, *Microwave News.*

Swicord, Mays, Ph.D.; FDA official, then Motorola employee.

Taflove, Allen, Ph.D.; Northwestern University; instrumental in developing the Stealth bomber technology; helped create new exposure systems for WTR.

Tice, Raymond, Ph.D.; Integrated Laboratory Systems, Research Triangle Park, N. C.; researcher who, with Dr. Graham Hook, found first evidence that cell phone radiation causes DNA breaks.

Wheeler, Thomas E.; President, CTIA.

*Member of Peer Review Board

† Member of U.S. government Interagency Working Group on Radio Frequency Radiation

(A complete list of members of these panels may be found in the Appendix section)

APPENDIX

UNITED STATES GOVERNMENT INTERAGENCY WORKING GROUP ON RADIO FREQUENCY RADIATION

FOOD AND DRUG ADMINISTRATION:

Elizabeth D. Jacobson, Ph.D.
Deputy Director of Science
Center for Devices and Radiological Health

Russell D. Owen, Ph.D.
Chief, Radiation Biology Branch
Center for Devices and Radiological Health

Larry W. Cress, Ph.D.
Medical Officer, Radiation Biology Branch
Center for Devices and Radiological Health

Donald E. Marlowe
Director, Office of Science and Technology
Center for Devices and Radiological Health

Howard I. Bassen
Chief, Electrophysics Branch
Center for Devices and Radiological Health
Joanne Barron
Regulatory Officer, Office of Compliance
Larry G. Kessler, Sc.D.
Director, Office of Surveillance and Biometrics
Center for Devices and Radiological Health
Ronald G. Kaczmarek, M.D., M.P.H.
Medical Officer, Office of Surveillance and Biometrics
Center for Devices and Radiological Health

ENVIRONMENTAL PROTECTION AGENCY:
Joe Elder, Ph.D.
Special Assistant
National Health and Environmental Effects Research Laboratory
Norbert N. Hankin
Environmental Scientist
Office of Radiation and Indoor Air

NATIONAL INSTITUTE OF OCCUPATIONAL SAFETY AND HEALTH:
Gregory W. Lotz, Ph.D.
Chief, Physical Agents Effects Branch
Division of Biomedical and Behavorial Science

FEDERAL COMMUNICATIONS COMMISSION:
Robert F. Cleveland, Ph.D.
Physical Scientist
Office of Engineering and Technology

OCCUPATIONAL SAFETY AND HEALTH ADMINISTRATION:
Robert Curtis, Ph.D.
Director, Division of Program Support

DEPARTMENT OF COMMERCE:
Janet Healer, Ph.D.
Bioscientist
National Telecommunications Information Administration

NATIONAL INSTITUTE OF ENVIRONMENTAL HEALTH SCIENCES:
Mary Wolfe, Ph.D.
Associate Coordinator, EMF Hazards

NATIONAL INSTITUTES OF HEALTH:
Martha Linet, M.D.
Medical Officer, Radiation Epidemiology Branch
National Cancer Institute

PEER REVIEW BOARD
COORDINATED BY THE HARVARD CENTER FOR RISK ANALYSIS

Project Directors:

John D. Graham, Ph.D.
Director, Center for Risk Analysis
School of Public Health
Harvard University

Susan W. Putnam, Sc.D.
Research Associate in Environmental Policy
Center for Risk Analysis
School of Public Health
Harvard University

Members:

Larry E. Anderson, Ph.D.
Bioelectromagnetics Program Manager
Departments of Biology and Chemistry
Battelle Pacific Northwest Laboratories

Patricia Buffler, Ph.D., M.P.H.
Dean, School of Public Health
University of California at Berkeley
Sir Richard Doll, F.R.S., F.R.C.P.
Emeritus Professor of Medicine
Oxford University, England
Carl H. Durney, Ph.D.
Professor, Electrical Engineering Department
University of Utah
Joe Elder, Ph.D.
Special Assistant
National Health and Environmental Effects Research Laboratory
U.S. Environmental Protection Agency
Saxon Graham, Ph.D.
Professor Emeritus
State University of New York at Buffalo
Don R. Justesen, Ph.D.
Professor of Neuropsychology, School of Medicine
University of Kansas
Research Service, USVA Medical Center
Richard R. Monson, M.D., Sc.D.
Professor of Epidemiology, School of Public Health
Director, Educational Resource Center for Occupational Safety and Health
Harvard University
Asher R. Sheppard, Ph.D.
Consultant
Asher Sheppard Consulting
Dimitrios Trichopoulos, M.D., S.M.
Director, Harvard Center for Cancer Prevention
School of Public Health
Harvard University
Peter A. Valberg, Ph.D.
Cambridge Environmental, Inc.

Gary M. Williams, M.D.
    Chief, Division of Pathology and Toxicology
    Director of Medical Sciences
    American Health Foundation
Sheldon Wolff, Ph.D.
    Vice Chairman and Chief of Research
    Radiation Effects Research Foundation

ACKNOWLEDGMENTS

The work that led to this book would not have been possible without the help of Drs. Bill Guy, Ian Munro, and Don McRee. They were the scientific heart and soul of the WTR, and hung in there with me when the going was the toughest. They have my deepest appreciation.

This book itself would not have been possible without the fine work of our editor, Philip Turner of Carroll and Graf.

The Peer Review Board did an excellent job, and I owe a special debt of thanks to Drs. John Graham and Susan Putnam for all their hard work in keeping that part of the project going.

The Audit Committee, and especially Ron Cavill, did a great job overseeing the financial aspects of the WTR, and when the tensions were highest, they were very effective liaisons between me and the CTIA.

My legal advisors, Linda Solheim, Jim Baller, Bruce Navarro, and Mike Coffield, did a fine job counseling and defending, but more importantly they were my friends and helped me keep perspective.

Lisa Joson supported me and the project with everything she had. Her husband, Josh, and daughter, Jolie, sacrificed much by giving her the freedom to do so. Slavko Bradic, a former basketball star, was our WTR most valuable player day in and day out, cheerfully handling tasks large and small.

Becky Steffens-Jenrow oversaw the epidemiology and always added critically important scientific balance.

Jeff Nesbit was one of the few who was there at the beginning and at the end. His advice was always welcomed.

Kelly Sund, Thorne Auchter, and Cindy Perno did a great job coordinating the WTR staff at different times over the seven years of the project.

To the WTR staff, my personal thanks to all who helped: Dr. Jack Kille, Dr. Earle Nestmann, Marjan Najafi, Kathleen Kapetanovic, Martha Embrey, Susan Hersemann, Claudine M. Valmonte, Gretchen Findlay, Polly Thibodeau, Salem Fisseha, Abla Mawudeku, Sherry Farr, Jennifer Cohen, Courtney Kvancz, Marybelle Ang, Elizabeth Adams, Pete Sebeny, Sam Amirfar, Nancy Akers, Janis Lemke, Mary Supley, Sue O'Donnell, Stan Hankin, Bryan Eddins, Robert Bailey, Anita Sperling, Maureen Jablinske, Mike Niemeier, Niles Comer, Jody Dosberg, Patricia Carlo, Mike Volpe, Mary Beth Franklin, Don Jarvis, Dr. Bert Liston, Dr. John Vena, Dr. Elisa Bandera, and Collette Herrod.

I also thank the members of the U.S. government's Interagency Working Group for important scientific guidance.

And many thanks to the scientists who did the work—especially Dr. Ray Tice and Dr. Graham Hook. You have gotten more public visibility than you bargained for. Thank you.

George Carlo

• • •

Thanks to literary agent Ronald L. Goldfarb, whose concern about the safety of cell phones led him to introduce two men from different professions, who discovered shared values and imperatives and became co-authors. And a special thanks to Philip Turner, executive editor of Carroll & Graf, whose science education led him to appreciate the importance of this book, and whose literary skills and attention to detail assured that its message would reach the people who need it most of all.

Martin Schram

Index

X-Y

About the Authors

Dr. George Carlo is a Fellow of the American College of Epidemiology, and serves on the adjunct faculty of The George Washington University school of Medicine. Dr. Carlo has served in many scientific advisory capacities over the past 20 years, in government and industry. He has testified before Congress and other government and regulatory bodies in the United States, Europe, and the Pacific Rim, and has written more than 150 scientific papers, treatises, and chapters in books. He earned his masters and doctoral degrees from the State University of New York at Buffalo and his law degree from The George Washington University National Law Center. Dr. Carlo is presently chairman of Health Risk Management Group in Washington, D.C.

Martin Schram has been a journalist covering the workings of government and politics for more than three decades. His nationally syndicated newspaper column, which focuses on the intersection of the news media, policy and politics, is distributed nationally by the Scripps Howard News Service to more than 400 newspapers. He appears frequently on national television news programs. Schram has received a number of honors for his writings, including the 1988 Lowell Mellett Award for Outstanding Media Criticism for his book, *The Great American Video Game.* His other books include: *Speaking Freely: Former Members of Congress Talk About Money in Politics*, published in 1995; and *Running for President 1976: The Carter Campaign*, published in 1977. Schram received his B.A. with a major in political science at the University of Florida in 1964. He began his career in news at *The Miami News.* Schram joined *Newsday* as a reporter in 1965, and in 1972 was named

Newsday's Washington bureau chief and senior editor. He moved to the *Washington Post* in 1979, serving as the newspaper's writer on the presidency and as a national affairs correspondent. He and his wife, Patricia, live in the suburbs of Washington, D.C.